The GRC Exception Handbook: A Practical Guide for Modern Organizations

Glenn Haggard

Gruntworks Technology LLC

Contents

Dedication

F or every GRC, IT, and security professional who's ever felt alone at the intersection of business needs and audit checklists: this book is for you.

To my friends and family who kept me going, as well as to my past and present peers and colleagues.

But mostly to my wife, Gina, and my son, Glenn, whose support and belief shape everything I do. Thank you for inspiring me.

Introduction

H ave you ever found yourself scrambling to fix a policy gap, sweating through an audit, or wishing that you had a playbook when the rules stop making sense? If so, *The GRC Exception Handbook: A Practical Guide for Modern Organizations* could be precisely the resource you've been searching for.

As a cybersecurity specialist with more than a decade of experience navigating Governance, Risk, and Compliance in demanding environments, I've seen firsthand how well-managed exceptions can change the entire trajectory of a risk program. They can shift teams from reactive cleanup to proactive problem-solving that strengthens resilience. When they are handled poorly, they are often the root cause of costly failures and reputational damage.

This book distills those experiences into a practical toolkit that combines proven strategies, real-world examples, and steps you can apply immediately. Whether you are building a GRC program from the ground up or refining one already in place, you will find guidance that helps you manage exceptions with clarity and confidence.

Why GRC and Exception Management Matter Now

Ideally, every system and organization would be 100% compliant. In the real world, however, exceptions happen. Like a leaky faucet that floods if left unattended, situations will arise. With remote work and AI expanding risk surfaces, the *2025 Verizon Data Breach Investi-*

gations Report found that humans were a factor in roughly 60% of breaches, often due to unaddressed gaps.

Ignoring proactive Issues and Exceptions Management (IEM) can lead to severe compliance failures, crippling financial penalties, and long-term reputational harm. Mastering IEM can be the difference from reactive firefighting to proactive risk mitigation, building resilience, driving efficiency, and supporting innovation.

Who This Book Is For

This book is designed for anyone who wants to strengthen organizational defenses through effective IEM. Whether you're a security specialist, IT manager, compliance officer, or business leader seeking practical answers, this book is a great resource. No prior deep GRC expertise is required.

Each concept is explained clearly and grounded in a real-world context, not abstract theory. Whether you're in a startup defining its first controls, or a large enterprise navigating complex frameworks like GDPR (Europe's data privacy law), SOX (U.S. financial reporting law), or CMMC (the Department of Defense's cybersecurity standard for contractors), you'll find insights adaptable to your environment.

How to Use This Book

Before diving in, it helps to understand how this guide is structured. Think of it as your **GPS for GRC**. You can read from cover-to-cover, or jump directly to the chapter you need, with signposts and cross-references to keep you oriented.

To enhance learning and application, chapters include recurring components such as:

- **If You Only Remember One Thing:** A concise summary of each chapter's key takeaway, highlighting the core concept of that chapter.

- **Quick Wins:** Actionable steps that you can put in place immediately.

- **Practical Examples and Case Studies:** Real-world stories illustrating the concepts in action.

- **What Not to Do:** Common pitfalls and lessons learned. Learning from your mistakes is beneficial; learning from the mistakes of others is even more so.

- **Manager's Perspectives and Practitioner's Notes:** Tailored insights for different audiences.

The Overall Flow of the Book

This book follows a deliberate arc, similar to the way a consultant guides an organization through a complex transformation. We begin by setting the stage in the Introduction and Context chapters, then move through the practical work of lifecycle design, remediation, documentation, and reporting. We end by looking ahead to future challenges and long-term improvement.

The overall structure moves the reader from awareness to application to mastery. We start by explaining why exceptions matter, then show what effective management looks like, and finally explore how to sustain and refine a mature program.

Each chapter mirrors the consulting arc I use with clients: frame the challenge, teach the principle, troubleshoot the real-world mess, and inspire improvement. This pattern helps turn theory into practice and practice into progress.

Key Terms and Conventions

Early in my career, I learned that jargon can hinder effective communication. I don't want that to happen here. To ensure clarity, especially for those new to the intricacies of GRC and IEM, key terms are defined as they appear and standardized throughout the document. A glossary is included at the end of the book for quick reference.

Throughout the text, *italics* highlight emphasis, **bold** marks new terms or definitions, and monospace font identifies templates or sample text for easy use. Acronyms are spelled out on first use.

Diversity of Examples: Insights from Finance, Healthcare, Tech, and More

One of the core strengths of this book is its broad applicability across sectors, reflecting lessons learned in real programs I've led. The best GRC solutions often draw inspiration from diverse fields:

- **Finance:** How banks manage exceptions in anti-money laundering (AML) controls to avoid multimillion-dollar fines.

- **Healthcare:** Where HIPAA (Health Insurance Portability and Accountability Act of 1996) compliance lapses can compromise patient trust and invite severe penalties.

- **Technology:** How cloud, SaaS, and AI providers balance rapid innovation with vendor and configuration risks.

- **Defense Contractors:** How organizations pursuing or maintaining CMMC (Cybersecurity Maturity Model Certification) navigate exception handling, POA&M (Plan of Action and Milestones) management, and evidence collection across complex supply chains.

We'll also explore manufacturing, retail, education, and government to illustrate universal principles, such as utilizing automated workflows to track exceptions and maintain accountability. Regardless of the industry, our goal remains the same: to transform exceptions from compliance burdens into catalysts for performance and trust.

Let's dive in. By the end, you'll turn GRC from a headache into your secret weapon for organizational resilience. Chapter 1 sets the stage by putting GRC and exception management in a real-world context. Using easily understandable examples enables us to clarify what *issues* and *exceptions* are, why they matter, and how to define a simple mission statement to guide your team.

Chapter 1: GRC and Exception Management in Context

If You Only Remember One Thing: GRC Balances Opportunity with Accountability.

Have you ever faced a deadline so tight you had to bend a rule to get the job done? Perhaps it was a product launch, a system outage, or a new regulation that forced a quick decision. These moments test policies and controls to their limit. That's where smart Issues and Exceptions Management (IEM) makes the difference between resilience and regret.

As promised in the introduction, this first chapter frames what Governance, Risk, and Compliance (GRC) really means in context, and why it exists. Additionally, we will examine how an effective IEM process can keep organizations accountable and why getting it right is essential.

GRC is how organizations grow responsibly. It ensures innovation doesn't outpace control, and that growth doesn't come at the expense of compliance. Let's explore GRC and take it from a buzzword to a competency.

What is GRC?

Managing GRC is like guiding a ship. You need a clear destination (governed by your business goals), a map to avoid hidden reefs (risk management), and a crew that adheres to maritime laws (compliance).

- **Governance** establishes the rules and clarifies ownership and accountability. Whether it's a board approving new policies or a manager delegating authority, good governance means everyone knows the playbook and the score. *Why it matters: Governance prevents chaos. When everyone knows who's steering and why, the ship stays on course even in rough waters.*

- **Risk Management** identifies and addresses threats before they escalate into disasters. Sometimes that's as technical as prioritizing vulnerabilities after a scan, or as strategic as planning for supply chain interruptions. *Why it matters: Risk management helps you avoid costly surprises by spotting the iceberg before you hit it, not just patching leaks after the fact.*

- **Compliance** ensures adherence to external laws, such as HIPAA (the Health Insurance Portability and Accountability Act, a U.S. healthcare privacy law), and GDPR (the General Data Protection Regulation, an EU privacy law), as well as internal standards established by leadership. It's how you demonstrate to regulators, auditors, and customers that your business is trustworthy. *Why it matters: Compliance supports trust and keeps doors open. No one wants to do business with a company one audit away from disaster.*

GRC isn't a place reserved for just lawyers and techies. It's every leader's responsibility. In a startup, it may mean protecting customer data while growing; in a global enterprise, it could involve aligning distributed teams with relevant privacy laws and regulations. More than just being a checklist, GRC is a mindset. It can enable your company to pursue ambitious goals while maintaining a vigilant approach to risk and accountability.

Takeaway: GRC isn't just one team. It's every leader's toolkit for ambitious growth with eyes wide open to risks.

Why Exception Management is Core to GRC

Organizations write policies to create stability, but real life rarely follows the script. How your organization handles exceptions determines whether it adapts successfully or collapses under pressure. The best GRC programs know when to approve exceptions and when to say no.

Exception management is generally the term teams use to refer to IEM in general, and it does not imply a lack of issues. IEM is a structured process for handling situations when standard controls are not applicable. When done right, it keeps your business flexible and resilient. When done poorly, it invites chaos, audit findings, and even front-page scandals.

Compliance requirements differ across industries. The defense sector is now using the Cybersecurity Maturity Model Certification (CMMC), for example. This framework expects defense contractors to document and review any approved security exceptions. This expectation is detailed in practice RM.L2-3.11.2 and requires a written record of known risks and their ongoing management. As this shows, exception management isn't only innovative governance; it is also a formal compliance obligation.

Examples in action:

- **Urgent needs:** A sales team may request an exception to skip a vendor review. This type of exception can become problematic if not followed up on and documented thoroughly.

- **Resource shortages:** In a hospital, budget cuts could delay a compliant MFA rollout. The downstream effects could be enormous. Document the gap to spark upgrades.

- **System limitations:** Legacy software (and hardware) can be problematic even in newer companies. Document it to avoid audit findings.

- **Human error:** Mistakes happen. Amazon had major outages in March 2017 and again in October 2025, which they attributed to human error. Quick fixes may resolve the issue, but logging them can help to prevent recurrence.

- **Regulatory change:** New frameworks or court rulings can necessitate adjustments in how organizations operate. IEM teams should document the changes and maintain a clear timeline to ensure compliance.

Did you notice that documenting what happened and why was a core part of all those examples? It is a significant part of IEM, but we will explore that in later chapters.

Imagine a team needs to implement a vendor's tool that doesn't yet meet security standards. This situation is where an exception would come in. With clear documentation, an expiration date, and assigned responsibility, the team could potentially move forward safely. Skip those steps, however, and the risk becomes liability.

Exceptions aren't always a bad thing. They shouldn't be treated as such, either. When handled right, they can spark practical solutions. I saw this firsthand working at a large multinational software company where I led a team that managed high-profile IEM escalations.

When leadership suddenly banned all contingent worker access, the fallout was immediate. In only a few days, projects were stalled and tickets piled up. Instead of issuing blanket waivers, our IEM team collaborated with each affected business unit and key stakeholders to design specific, risk-based exceptions with proper guardrails in place.

The result was that critical projects resumed quickly and securely, with leadership gaining new trust in the IEM process. By refusing to rubber-stamp exceptions, we avoided costly outages, maintained auditability, and earned the trust of our leadership. The collaboration on this matter helped spur our organization to refine the process for changing policies, which was instrumental in ensuring that sweeping changes underwent proper review from a sufficient number of stakeholders.

A well-designed exception process can turn chaos into clarity and help organizations stay both agile and secure. Issues and Exceptions Management can be a catalyst for positive change within your organization, enabling more effective business processes and improved results.

Remember: If your IEM team does not track, review, and ensure exceptions are time-limited, you are not managing risk; you are deferring it.

Practitioner's Note: Where Exception Management Fails

Every exception should be tracked in a central GRC system, reviewed by both business and risk stakeholders, and set to expire automatically. Any "permanent exception" is a future audit finding in waiting, and a clear sign of an immature IEM process. In that chaos over contingent worker access, refusing blanket waivers built trust and proved that when exceptions are time-limited and monitored, they aren't roadblocks but accelerators.

Tie-in: The same principle applies whether you're managing vendors, systems, or regulatory deadlines. Measured exceptions build credibility, while unmanaged ones erode it. The remainder of this chapter illustrates where exceptions occur in your organization and how to prevent them from becoming silent risks.

What Not to Do: An Ounce of Prevention

- Never approve an exception informally or as a handshake deal. The risk that isn't documented is likely to be a finding in your next audit.

- Never let "temporary" become "permanent." As we often say in the industry: *There is nothing more permanent than a temporary solution.*

- Avoid blanket exceptions for an entire technology ("Cloud is different, so we'll revisit later..."). Each exception must be specific, time-limited, and have a remediation plan.

- Never skip involving both business and security stakeholders. Unilateral decisions create blind spots and undermine trust.

Bottom line: Treat every exception as if it will be scrutinized in your next audit, legal review, or customer inquiry. How you manage exceptions determines whether your organization can adapt safely or invite new risks.

The Actual Value of a GRC IEM Team

A strong IEM team does far more than satisfy compliance checklists; it functions as a true business partner. By tracking and resolving exceptions, the team acts as an early warning system for emerging risks. Examining repeated exceptions tied to the NIST Cybersecurity Framework (Identify, Protect, Detect, Respond, and Recover) will often reveal more profound, systemic weaknesses.

Recognizing those weak spots can help your team to justify targeted investments that align with SOC 2 criteria, increasing security, availability, and confidentiality. A vigilant IEM team can spot issues early, close compliance gaps, and prevent costly surprises that affect both reputation and operations.

Just as significantly, a mature GRC IEM function shifts an organization from reacting to problems toward building resilience. Each exception tells a story: when analyzed and corrected, it provides data that exposes root causes and guides better decisions.

For example, if a growing number of teams file exception cases with the same software tool, it could indicate outdated technology or a policy that no longer aligns with the business. Understanding this trend can help your IEM team provide leadership with a clearer view of where resources should be allocated, which controls require adjustments, and where training may be falling short. This steady cycle of observation, correction, and improvement keeps the organization learning, strengthening its controls, and staying ahead of changing risks.

Even if you're outside security, think of IEM as your organization's health check, catching minor problems before they grow into crises. As the saying goes, *an ounce of prevention is worth a pound of cure.* An IEM team is like your ship's lookout spotting icebergs early so you can steer clear, not just patch holes after the crash.

Takeaway: A strong IEM team turns policy gaps into learning opportunities and helps your team adopt continuous improvement, not just compliance.

Even with a strong GRC team, challenges don't always fit neatly into pre-made solutions. Sometimes what appears to be an exception is actually a deeper issue. A strong GRC team can make the difference between a managed exception and a significant finding.

Why Exceptions Happen

Why do exceptions arise, even in high-performing teams? Real-world constraints rarely fit the textbook. Even in well-run organizations, exceptions pop up like unexpected roadblocks on a cross-country drive. We can return to the example of urgent needs: A sales team pushes to skip contract review, hoping to close a deal before a rival. A 'just this once' exception that can set a precedent, like a shortcut turning into a habit if not mapped out.

Sometimes an exception stems from system limitations. Legacy software can carry known vulnerabilities that a team can't patch without risking disruption to core operations. In one retail chain where I worked, an aging inventory platform failed to meet PCI DSS requirements. The team had to implement a temporary workaround, which was tightly controlled and clearly scheduled for complete remediation.

Human error, such as a misconfigured firewall during a Black Friday rush, can undoubtedly lead to exceptions, as can a change in regulation, like the GDPR.

The most concerning cause of exceptions is a shortage of resources. Budget delays or staffing gaps shouldn't become routine excuses. Scarcity doesn't excuse poor planning.

An important concept to drive home, perhaps most critically to senior management, is that exceptions aren't failures. They are instead signs that the world is moving fast, and your GRC process must keep up. The key is to assess the risk, document the decision, and plan for a real fix. Understanding why exceptions happen is only half of the battle. The next step is recognizing where they surface inside your organization, and how to balance business urgency with compliance reality.

Exception Triggers Checklist

These five triggers account for the majority of exception requests I've seen in the field. Use this quick tool to identify them early and address the issue before it escalates. Categorizing them can be helpful when reviewing your case logs in the future.

[ ] Urgent Need

[ ] Resource Shortage

[ ] System Limit

[ ] Human Error

[ ] Regulatory Change

Log these to spot patterns early. Now that we know why exceptions happen, let's explore where they show up in your business.

Where Exceptions Fit: The Intersection of Business, Risk, and Compliance

Every exception sits at an intersection of business need, risk tolerance, and compliance. These three competing interests can have very different outcomes in mind for the exception. Managing this tension is what separates mature organizations from those stuck in perpetual "fire drill" mode.

I'll never forget the time a team at a biotech startup handled an urgent system outage by bypassing standard change controls in the name of "speed." They didn't document the changes made, and they didn't inform management. They believed they had saved the day, since the fix worked. *Temporarily*.

When auditors later reviewed the incident, they noted the undocumented exception as a significant finding. They had to scramble to reconstruct decisions made months earlier, and their memory proved to be a poor recordkeeping system. The "fire drill" approach ultimately costs more in reputation and rework than it would have to document the exception properly.

These pressures usually show up in three predictable ways, each pulling the organization in a different direction.

- **Business Need:** A marketing team wants to launch a new campaign using a tool that has not yet cleared security review. Approving the exception might mean a revenue win, but at what risk?

- **Risk Management:** Sometimes, granting a vendor temporary access is necessary for a project, but only if the proper safeguards are in place.

- **Compliance:** System outages that delay regulatory reporting require documentation and a clear plan to get back on track.

Tech Example: At a SaaS startup I worked with, a product team fast-tracked a connection to a partner's software. A senior VP insisted that the exception bypass the standard due diligence checklist over objections from the security team. When a vulnerability was later discovered in the partner's system, our company faced tough questions from customers, auditors, and certifying bodies. Skipping the process that the governance team had put in place to protect the company left us exposed.

Takeaway: Every exception is a negotiation between what the business wants, what's risky, and what's allowed. Success means managing all three.

Emerging Risks: How Technology Shapes GRC Challenges

Technology accelerates both opportunity and risk. Artificial Intelligence (AI), cloud computing, and automation change how exceptions arise and must be managed.

- **AI and Automation:** At a previous employer, we piloted an AI-powered analytics tool that needed broader data access than our policy allowed. We approved a temporary exception, but only after documenting privacy risks, restricting access, and scheduling a review. Even so, we learned that AI tools like chatbots can easily outpace policy controls, leading to unanticipated risks if not managed closely.

- **Cloud Systems:** Moving to cloud platforms can make security policies obsolete overnight. A client once needed an exception for a cloud file-sharing tool that didn't meet his company's encryption standards. The team was granted the exception, subject to strict monitoring and a timeline for compliance. With enough time for the vendor to upgrade to a more rigorous encryption standard, the team was able to move forward. For a retail manager new to GRC, cloud misconfigurations (like skipping encryption) can seem minor until they trigger PCI DSS fines. It can become critically important to document exceptions with safeguards to turn threats into secure shortcuts.

- **Cybersecurity Threats:** Urgent cyber threats or vulnerabilities sometimes force organizations to operate "out of policy" to stay secure. The best-prepared organizations are those that plan for these exceptions in advance, documenting procedures and not improvising solutions under pressure.

- **Quantum Computing:** A future technology that could render current encryption vulnerable, demanding early strategic planning for your data and systems. (See Tech Deep Dive for more).

AI Example: In 2014, Amazon trained an AI recruiting tool on a decade of resumes. Within a year, bias against women became apparent, and Amazon soon scrapped the tool. This incident was an operational and reputational failure that highlights the need for governance to evolve in tandem with innovation.

Cloud Example: A small business used a cloud storage tool that didn't meet encryption standards, requiring an exception. Without proper monitoring, a simple misconfiguration led to a data leak and $100,000 in fines. They could've prevented all of that with a documented exception showing safeguards in place.

I once worked with a team racing to deploy a cloud tool before a major product launch. We granted an exception but did not monitor it. When a breach hit, it felt like a story where the hero ignores a warning. It was costly but avoidable with a better process.

Tech Deep Dive: Quantum Computing's GRC Implications and What to Do Now

Quantum computing may eventually render current encryption obsolete. Forward-looking organizations already track NIST's post-quantum algorithm standards and assess vendor readiness. The goal isn't panic, it's preparation.

Why it matters to GRC:

- Current encryption could be rendered obsolete overnight once quantum attacks capable of breaking asymmetric encryption are viable, threatening everything from secure transactions to stored archives.

- Privacy risks will increase as anonymized data becomes easier to re-identify users from currently anonymized data.

- AI governance may become more challenging as quantum-scale processing enables faster and subtler algorithmic bias.

- New supply chain risks will emerge as quantum components mature.

What to do now:

1. **Inventory sensitive data** that could be at risk from future quantum encryption attacks and **monitor NIST's evolving post-quantum cryptography standards** (e.g., ML-KEM (for encryption), ML-DSA and SLH-DSA (for digital signatures), and HQC (a code-based algorithm under evaluation by NIST for future PQC standards). Track these as a long-term IEM risk.

2. **Engage key stakeholders** in security, compliance, and legal to assess contractual or regulatory impacts.

3. **Evaluate vendor readiness,** especially for cloud providers and critical third parties, to ensure they have a post-quantum roadmap.

The point isn't to panic. It's to document your awareness, track the risk, and be ready to pivot when post-quantum standards become operational. That's how GRC leaders stay credible in the face of disruptive change.

The Real-World Stakes: Short, Relevant Case Studies

Sometimes, failure to manage exceptions can make headlines or sink companies altogether.

Healthcare Example: During the COVID-19 pandemic, hospitals had to pivot quickly. At the EMR (Electronic Medical Record) company where I once consulted, leadership granted a temporary exception to allow a non-compliant telehealth tool that was essential for seeing patients remotely but lacked full encryption.

Because we documented the decision, kept it time-limited, and flagged it for review, the company met the needs of the medical providers who used their product and avoided regulatory trouble.

Why: Without documentation, this could have led to a breach and a $500,000 fine.

Banking Example: In 2016, Wells Fargo's sales teams created millions of unauthorized accounts to meet aggressive performance goals. Weak oversight and a lack of accountability allowed control violations to go unaddressed. This became a textbook example of what not to do, as it ultimately led to fines exceeding $3 billion and the loss of both reputation and executive leadership.

Root Cause: A culture that normalized policy exceptions and tolerated rule-bending under pressure led to systemic control failures, unchecked compliance risks, and widespread misconduct.

Manufacturing Example: In their 2018 rush to launch the 737 MAX, Boeing overlooked safety exceptions, leading to crashes, regulatory scrutiny, and billions in losses, as well as the tragic loss of 346 lives.

Why: The core failure was in overlooking exceptions to safety protocols and allowing urgent business priorities to override engineering judgment. Without a structured review, this ultimately resulted in catastrophic consequences.

Retail Scenario: A family-owned store opted not to implement encryption to rush its online launch. A data breach ensued, resulting in six-figure fines and irate customers. A simple, well-managed exception process could have prevented it.

Turning Point: The absence of a formal exception process turned a simple shortcut into a nightmare; proper documentation could have managed the risk and sidestepped the breach, six-figure fines, and furious customers.

Each of these failures shares a common root cause: unmanaged exceptions. The following section demonstrates how a clear and simple GRC mission statement can prevent such missteps by defining accountability before a crisis occurs.

Early in my career, I watched a team bypass a security check to meet a deadline, like characters in a thriller ignoring a warning sign. The exception wasn't documented, and a minor breach cost us weeks of cleanup. That taught me that every exception needs a story with a clear beginning, middle, and end, all of which are documented and resolved.

Whether you're running a hospital, a bank, a tech company, or a family business, exception management isn't just about compliance. It's about protecting your mission, your people, and your brand.

A robust GRC program is an investment with immense ROI. If Boeing or Bank of America had spent millions annually on GRC, the prevention of these scandals would have paid for itself many times over.

Quick Win: Draft a One-Page GRC Mission Statement

I've seen organizations transform their GRC culture with a clear mission statement. The most effective mission statements aren't just written, they're lived. Bring your GRC mission into onboarding, team meetings, and policy reviews. Fill this in like a Mad Lib. It'll spark discussions that catch exceptions early, as it did in my first role.

How to Create Yours:

1. **State Your Purpose:** "Our company thrives by pursuing growth while staying secure and compliant."

2. **Define Key Principles:** Examples: "We prioritize customer trust," or even "We manage risks proactively."

3. **Address Exceptions:**"We manage exceptions with transparency and a remediation plan."

4. **Make It Actionable:**"Every team member is responsible for upholding these standards."

5. **Keep It Simple:**Keep it on one page and write it in plain language. The above suggestions are just starting points.

Share your new, unique GRC mission statement in the new employee onboarding. Don't let it be seen once and then just forgotten. Discuss it in team meetings and revisit it whenever your business or compliance environment changes. When everyone knows the "why" behind GRC policies, you'll see earlier exception requests, better documentation, and fewer last-minute crises. Don't just write it, live it,

In my first InfoSec role, a mission statement posted in our office sparked a team discussion that caught an exception early, saving us from an audit failure. It showed me that a clear mission statement isn't just a collection of words; it's a rallying cry.

Manager's Perspective:

Exception management isn't just a compliance requirement. It's risk management in action. As a leader, your team will take the process seriously only if you do. You can set the tone by reviewing exception reports, not just the outcomes.

Ask: "What exceptions have we approved this quarter? Are any still open, and why?" This visibility prevents a culture of silent workarounds. Such visibility is the cornerstone of transparent risk management. Like the firewall scramble I mentioned earlier in the chapter, request quarterly reports to turn silent risks into open dialogues.

The Promise and Roadmap of This Book

I wrote this book to be your practical guide to mastering GRC and Issues and Exceptions Management. Whether you're a business owner, manager, or new to GRC, you'll find actionable tools, examples, and strategies to strengthen your organization.

Here's what's ahead:

- **Chapter 2: The Language of Risk: Exceptions vs. Issues**We'll break down how exceptions differ from issues, using clear definitions, decision frameworks, and real-world examples.

- **Chapter 3: Designing and Managing the Exception Lifecycle**Follow the whole journey of an exception, from intake to closure, with practical templates, workflows, and automation tips.

- **Chapter 4: Exception Requests and Remediation**Learn how to evaluate requests, prioritize fixes, and turn exceptions into opportunities for stronger processes.

- **Chapter 5: Documentation, Audit, and Preventing Exception Creep**See how to document effectively, pass audits, and avoid pitfalls such as permanent exceptions and technical debt.

- **Chapter 6: Reporting, Dashboards, and Creating a Culture of Accountability**Explore how to design dashboards that tell a story, strengthen trust, and reduce exceptions through education and accountability.

- **Chapter 7: The Path Forward: Next Generation Risks and Continuous Improvement**

Prepare for what's next (AI, cloud, and beyond) with a checklist to keep your GRC program adaptable and resilient.

- **Chapter 8: FAQ/Troubleshooting: Navigating GRC Exceptions Like a Pro**

A quick reference for everyday challenges, lessons learned, and practical fixes.

- **Chapter 9: Deep Dives: GRC Battles Won and Lost. Practical Lessons from the Trenches**

Learn from past experiences and refine your plans.

By the end, you'll have a playbook to confidently manage exceptions, making your organization more resilient and trusted.

Small Team Tip: Don't wait for a "big" program to get started. Even a single-person GRC team can draft a mission statement and set priorities. Clarity attracts buy-in and resources. Focus on documenting *one* real business risk or exception per month to build momentum.

Like wrapping up a chapter in a novel, here's your action plan to follow and build momentum. Before moving to the next chapter, confirm you can check off each of these steps to keep exceptions from becoming permanent liabilities.

End-of-Chapter Checklist

- Draft your team's GRC/exception management mission statement.

- Identify at least one recent business scenario where exception management added value (or should have).

- List the main stakeholders in your GRC process (look at business, governance, risk, compliance, etc.).

- Schedule a 30-minute session to discuss GRC priorities with your leadership team.

- Review one emerging technology or trend that might impact your GRC efforts this year.

Next, we'll tackle one of the most common sources of confusion: how to tell an *exception* from an *issue*. Understanding the difference is the foundation of effective risk communication.

Chapter 2: The Language of Risk: Exceptions vs. Issues

If You Only Remember One Thing: Exceptions are Temporary; Issues are Systemic.

E ver driven with a warning light blinking at you? If it's a one-off glitch, just a sensor acting up, you fix it and move on. That's like an *exception*: a controlled, temporary deviation. But if that light keeps flashing because the engine itself is failing, that's an *issue*: a deeper, systemic problem.

In GRC, understanding the distinction between the two is crucial. Get it wrong and you can invite chaos, failed audits, or worse. This chapter explains how to distinguish between them, provides tools to evaluate each, and demonstrates how to maintain consistency and credibility in your program.

Definitions and Practical Distinctions

In GRC, clear language keeps your team aligned and your business safe. Exceptions and issues may appear similar, but their stakes and solutions are vastly different.

Exceptions are intentional, time-limited departures from your policies, standards, or controls. Think of them as detours. If you know a road is closed before you take a trip, you alter your plan. With an exception, you make a plan, document it, and ensure you get back on track.

Pro Tip: Teams under tight Service Level Agreements (SLAs) sometimes request exceptions when their deadlines are at risk of being missed. Resist the pressure to use the IEM process to bypass security or compliance controls. Do not allow the IEM process to become a fallback for missed deadlines. When an exception is necessary, you must pair it with a compensating control (a temporary safeguard to reduce the risk) and limit its duration. For instance, if you temporarily turn off a firewall rule (exception), you might add extra monitoring or network segmentation (a compensating control) during that brief window.

Understanding the types of controls involved can further clarify the distinction. GRC controls broadly fall into three categories: technical, administrative (also referred to as managerial), and physical. Each carries unique risks when exceptions arise.

- **Technical Controls:** These are safeguards embedded within technology systems. Examples are encryption settings, firewall rules, access permissions, multi-factor authentication (MFA), or automated vulnerability scanners. An exception here might be temporarily turning off MFA for a specific task. An issue would be a misconfigured service that did not require MFA.

Technical controls are like a car's airbags, automatic safeguards that protect you.

- **Administrative Controls:** These are policies, procedures, standards, and training programs that guide human behavior and define responsibilities. Your company's data retention policy, mandatory security awareness training, or a vendor selection process are examples of administrative controls. An exception might be a senior executive temporarily bypassing a standard approval process for a critical hire. An issue would be a consistent failure of multiple employees to complete mandatory training modules.

Administrative controls serve as the driver's manual, guiding behavior through established rules and regulations.

- **Physical Controls:** These are measures designed to protect physical access to assets. These controls include locked doors, security guards, surveillance cameras, alarm systems, and access badges for the data center. An exception might be granting temporary, escorted access to an unvetted contractor to a secure server room for urgent maintenance. An issue would be finding an unsecured server room door left ajar regularly.

Physical controls include locked doors that bar entry.

When reviewing an exception request, start by identifying the type of control that is affected. A gap in a technical control creates more immediate risk, so it usually requires stronger compensating safeguards. A failure in an administrative control often points to a broader issue with policy enforcement or training. Understanding the difference helps you identify the source of the deviation and its potential impact on the organization.

Examples:

- Skipping a security review to patch a vulnerability that attackers could exploit.

- Allowing a vendor short-term access to a sensitive system for a time-bound project.

- A legacy system, set to be replaced in three months, can't support a 22-character password, which violates corporate policy for administrative accounts.

Issues are systemic problems, often uncovered by audits or incidents, that expose more profound weaknesses. They're like a crack in the engine block. Ignore them and you risk catastrophic damage.

Examples:

- Repeatedly failing to encrypt customer data, which would violate SOC 2, a data protection standard for tech and SaaS companies.

- A misconfigured server, flagged by an ISO 27001 audit (the global standard for Information Security Management Systems), is exposing sensitive information to the public.

- Violating PCI DSS, a standard for securing credit card data, by storing Sensitive Authentication Data (SAD) after authorization.

Sorting exceptions from issues is like separating a sprain from a fracture. Both hurt, but one needs a quick wrap, the other a cast. Early in my career, a team mislabeled recurring vendor gaps as exceptions, leading to audit hits. We learned from that experience quickly.

Key Difference: Exceptions are deliberate, temporary, and backed by a resolution plan. Issues are patterns, often unintentional, revealing gaps in controls or compliance. Mistaking one for the other can lead to fines.

Practitioner's Note: Know Your Frameworks & Train Your Team

Auditors don't just look for paperwork; they look for patterns.

- **SOC 2** is your benchmark for customer data privacy in tech and SaaS companies.

- **ISO 27001** sets global standards for securing sensitive information.

- **The NIST Cybersecurity Framework (CSF)** is a risk-based guide for managing and mitigating cybersecurity risks. Note: In 2024, CSF expanded beyond its focus on critical infrastructure.

- **The General Data Protection Regulation (GDPR)** is an EU regulation that mandates strict rules for the collection, use, and protection of personal data.

- **The PCI Data Security Standard (PCI DSS)** is a mandatory standard for entities that process, store, or transmit credit card data.

- **Cybersecurity Maturity Model Certification (CMMC)** is primarily for defense contractors. In CMMC programs, systemic issues are documented as POA&M items that must be remediated within defined timelines. Temporary exceptions require formal risk acceptance and documentation at the same governance level as risk acceptance.

Remember: You can't win the GRC game if you don't know the rules, or if your team can't play by them. In a past role, ignoring NIST CSF patterns led to minor exceptions becoming significant issues. Train early and spot them.

Train your team to recognize and document differences. Use this ability to spot gaps before auditors do. With these definitions clear, let's dive into the high-stakes world of audits, where exceptions and issues tell very different stories.

Takeaway: Labeling a deviation correctly as a temporary exception or a deeper issue ensures you apply the proper fix and avoid future surprises.

With clear definitions in place, the next question is how auditors interpret them in practice, and how you can prepare.

Audit and Compliance Perspectives

Audits are not very different from doctor visits. When you visit the doctor, they will ask about your symptoms, examine your vital signs, and conduct some tests. With an auditor, you have interviews, evidence checks, and tests that reveal whether you're healthy or need major fixes.

Exceptions in Audits:

Auditors expect exceptions. They're a regular part of business agility, but the auditors will demand control. They expect to see documented risks, precise expiration dates, and a solid remediation plan.

Example: You approve a noncompliant cloud tool for a limited project. You should ensure that the exception is logged, justified by urgent need, and set to expire. Auditors will verify later that it was documented and that a remediation plan was implemented. If so, you pass.

Issues in Audits: These are red flags that require attention. They show recurring, unmanaged risks that can lead to failed certifications or hefty fines.

Example: An audit finds unencrypted customer data across multiple teams. It's not a one-time exception; it's a widespread issue, and it will cost you.

Why It Matters: Mislabeling an issue as an exception is like taping over a warning light. Eventually, the problem resurfaces, often in a worse form. Conversely, treating every exception as an issue can lead to unnecessary overhauls and slow down your business.

But audit findings are just the tip of the iceberg. The chill of mislabeling runs deeper.

Beyond Fines: The Deeper Implications of Mislabeling

While fines and failed certifications are tangible pains, the real cost of mislabeling or mishandling exceptions and issues runs deeper.

First, there's a false sense of security. If a persistent problem (an issue) is repeatedly swept under the rug as a "temporary exception," your organization operates under a dangerous illusion that its control environment is stronger than it actually is. Leadership may then allocate resources toward other threats, while real vulnerabilities remain unaddressed.

Second, it leads to ineffective resource allocation. Patching symptoms wastes time and budget on repeat fixes instead of solving the root problem. Imagine continuously buying Band-Aids for a leaking pipe instead of calling a plumber to fix the burst pipe behind the wall.

Third, it can erode internal trust and accountability. If teams see that legitimate issues are ignored, or that the GRC process is merely a "rubber stamp" for exceptions, they may stop reporting problems or take policies less seriously.

Eroded trust breeds complacency, which hides risk. If auditors spot this, it appears as a governance weakness. Their findings reflect the reality that the organization doesn't truly understand or control its own risks.

During an ISO 27001 audit, a fast-growing SaaS company encountered both sides of the exception-and-issue divide. First, auditors noted that the team's server log retention policy required 3 years of storage, whereas the older system could retain only 1 year. The GRC team treated it as a proper exception: they documented the gap, secured leadership approval, and laid out a straightforward plan to upgrade the system. The auditors accepted the approach and listed it only as an observation.

The same audit uncovered something very different. About 15 percent of employees had not completed mandatory security awareness training. Since the lapse had not been documented or tracked, and there was no clear plan to fix it, the auditors treated it as a compliance failure. One gap resulted in a minor note as an observation. The other, however, blocked a portion of the certification effort.

I worked with a team that documented a one-off exception for a cloud tool, thinking it was enough. That cloud exception became an issue in a SOC 2 audit, resulting in over $60,000 in fixes. A quick checklist would've caught it early, like spotting a plot twist before it derails the story.

Decision Framework: Exception or Issue?

Use this quick decision framework to stay agile **and** audit-ready:

1. **Is It Temporary?**

   ○ Precise end date, resolution plan? → Exception.

   ○ Keeps recurring with no plan? → Issue.

2. **Is It Intentional?**

   ○ Deliberate choice, with risks and approvals? → Exception.

   ○ Discovered accidentally, no record? → Issue.

3. **Does It Violate a Framework?**

   ○ One-time deviation, no breach of SOC 2/ISO 27001? → Exception.

   ○ Risks certification or legal standing? → Issue.

4. **What's the Scope?**

   ○ Affects a single project or team? → Exception.

   ○ Broad, organization-wide? → Issue.

Example: A marketing team skips a single security review, risking SOC 2 violation. If documented and limited, it's an exception; if routine, it's an issue that requires ISO 27001-level fixes.

The Power of Defaulting to "Issue"

The rule of 'when in doubt, treat it as an issue' isn't about excessive caution; it's about being prudent and proactive. When you default to "issue," you compel a more rigorous review. This issue designation will then automatically trigger a deeper dive into its root cause and demand a definitive long-term resolution. Issues often involve higher resource allocation, and greater leadership visibility. This mindset acts as a safety net, preventing minor one-offs from quietly turning into systemic vulnerabilities that compound over time.

Think of it as a quality control process in manufacturing. A recurring defect in a batch isn't a "minor acceptable deviation" (exception) but an "unacceptable defect" (issue) that signals a problem in the production line and requires investigation and a fix to the process. This strict interpretation prevents a critical systemic flaw from being missed, ensuring consistent quality and preventing recalls.

Once you've applied the basic framework, add one more layer of scrutiny: risk assessment.

Integrating Risk Assessment into the Decision

Risk assessment supports the exception-versus-issue decision, but it should not override the core criteria. Once you confirm that a deviation is temporary, intentional, and scoped, the next step is to assess whether the related risk remains acceptable. Temporary deviations can carry significant inherent risk, and compensating controls do not always reduce that risk enough to make the exception safe.

An example could be a request to pause the enforcement of a critical security control on a production server. Even if the request is intentional and time-bound, the underlying risk may be too high to accept. The temporary label does not soften the impact if something goes wrong.

A solid decision framework requires weighing the business benefit against both the potential impact and the likelihood that the risk will materialize during the exception period. If the risk remains unacceptably high, the appropriate course of action is to deny the request or classify it as an issue requiring immediate remediation. The goal is not to be restrictive but to ensure temporary gaps do not expose the organization to unnecessary harm.

What Not to Do: Common Classification Pitfalls

- Don't rely on verbal agreements. If it isn't documented, it doesn't count.

- Don't let "temporary" turn into "permanent." It's like allowing a guest to overstay until they become a squatter. Review exceptions before they become issues.

- Don't ignore scope. A one-time fix is fine, but a pattern is a red flag.

- Don't let exceptions pile up without a clear review schedule *.If you treat every problem as a one-off, you're just building tomorrow's audit findings. That's how minor oversights become systemic risk.*

Bottom line: If it's not documented comprehensively, for auditors and stakeholders alike, it simply didn't happen.

With the decision framework in hand, let's build a simple tool to make spotting exceptions and issues second nature.

Quick Win: Create a Simple Exception vs. Issue Checklist

Get it right every time by creating a one-page checklist to distinguish between exceptions and issues.

1. **Define the Problem**: Write down the deviation (e.g., "Using a noncompliant cloud tool").

2. **Ask the Framework Questions**:

- Is it temporary with a resolution plan?

- Is it intentional and documented?

- Does it comply with an external framework such as SOC 2 or ISO 27001?

- Is it limited to a defined scope?

1. **Document the Decision**: Both exceptions and issues require proper documentation, including owner and stakeholder approvals, along with a time-bound plan.

2. **Review Regularly**: Check exceptions monthly and follow up on remediation plans as needed. A lack of follow-up allows issues to roll over from one sprint into the next.

In my first InfoSec role, a checklist posted in our office sparked a team discussion that caught a mislabeled issue early, saving us from an audit hit. It felt like spotting the clue in time, proof that a simple tool can change the ending.

How to Use It: Share this checklist at project meetings, keep it posted in your GRC system, and review exceptions monthly to ensure they don't become issues quietly.

Even the best checklist fails without ownership. Before diving into real-world examples, map every exception to its framework and stakeholder.

Frameworks and Stakeholders

You don't have to memorize every regulation, but you absolutely must know which ones affect your business. Ensure your team has clear ownership for mapping exceptions to the relevant frameworks, SOC 2, ISO 27001, HIPAA, and PCI DSS. When a request arises, ensure your team considers which regulations or policies it impacts, and that all affected stakeholders are informed. Empower your team to push for clarity. Most audit headaches come from missed requirements or forgotten players. Frameworks are only as strong as the teams that apply them together.

Stakeholder Touchpoints for Exceptions

Exceptions should never be approved in a vacuum. Every decision requires three perspectives: business, technical, and risk. These are the lenses through which to look before saying yes.

- **Business** – Does this align with our objectives and priorities?

- **Technical** – Can we implement it safely, reliably, and sustainably?

- **Risk** – What are the regulatory, reputational, and operational downsides?

Each perspective typically comes from people in specific roles. Before granting an exception, confirm you've heard from these representative voices:

- **Requester's Manager** – Validates business alignment.

- **Process Owner** – Understands workflows and dependencies.

- **Control Owner** – Knows the control and compliance implications.

- **Risk/Compliance Lead** – Evaluates exposure and regulatory fit.

- **Technical Lead or SME** – Confirms safe, feasible execution.

- **Affected Business Unit Rep** – Provides on-the-ground impact insight.

- **Vendor/Partner Contact** *(if applicable)* – Identifies third-party effects.

Reality Check: In practice, you won't always have every role available. However, never skip any of the three core perspectives (Business, Technical, and Risk). Missing even one viewpoint can turn a calculated risk into an unmanaged liability.

Now let's see how these principles play out across industries.

In-Context, Real-World Examples and Scenarios

Let's see how exceptions and issues play out in the real world, with stories that show the stakes and lessons.

1. **Banking Compliance Issue: TD Bank's Systemic AML Failures**U.S. actions against TD resulted in approximately $3 billion in penalties, a guilty plea to money-laundering conspiracy, and the appointment of an independent monitor to oversee a multi-year remediation. Regulators cited widespread monitoring gaps and ineffective alerts. These are precisely the kind of "temporary exceptions" and backlog compromises that, if unmanaged, harden into organizational issues. FinCEN called the penalty record-setting for a depository institution. *Leader takeaway: Treat AML alert tuning changes and backlog deferrals as documented, time-limited exceptions with owners, milestones, and board visibility, or risk headlines.*

2. **Retail Issue: Unencrypted Data**A retail chain repeatedly bypassed its encryption requirements, labeling it a temporary workaround to cut costs. The shortcut remained in place from quarter to quarter, and the gap eventually exposed 10,000 customer records. When the SOC 2 auditors reviewed the situation, they flagged it as a systemic failure, resulting in the company losing its certification. The breach also triggered a lawsuit, negative press, and a two-hundred-thousand-dollar fine. A well-managed exception process could have limited exposure and prevented most of the fallout.

Lesson: Don't let a "workaround" become the norm. Documentation and follow-up processes matter.

1. **Finance Exception: Payment App Under the Microscope**A regional bank allowed a non-compliant payment app for a major client. Their product team filed an exception that documented the risks, set a remediation date, and secured approvals. The exception supported a $10M client deal, and that team's clear records saved them in a SOC 2 audit. A competitor's undocumented workaround led to a $100,000 fine and a loss of client trust.*Key Insight: Diligence in documentation can become a competitive advantage.*

2. **Healthcare Issue: EMR System Delays**A hospital repeatedly delayed updating its EMR system. They treated each delay as a separate exception. A HIPAA compliance review flagged it as a systemic issue, costing $300,000 in fines and eroding patient trust. Proper documentation and a remediation plan could have limited the damage.*GRC Gem: Patterns without resolution always become issues.*

Takeaway: Utilize the proper documentation process to transform a high-risk operational necessity (Exception) into a short-term, compliant detour, thereby preventing it from escalating into a costly systemic flaw (Issue).

Understanding what went wrong is only half the lesson. Here's how managers can prevent history from repeating.

Manager's Perspective:

As a manager, you're on the front line, making judgment calls that shape your business and reputation. Catching and fixing a problem before an auditor or customer does builds trust and credibility.

Think like a storyteller: every exception or issue has a beginning (why it happened), a middle (how it's handled), and an end (the resolution). When a sales team pressured me to bypass a contract review to close a deal, I insisted on documentation, set a 30-day review period, and involved the security team. We stayed agile and audit-proof.

Early in my career, I misjudged a recurring firewall misconfiguration as "just more exception." After the third round, an ISO 27001 audit flagged it, teaching me: if it repeats, it's an issue. Don't let speed cloud your judgment.

Balance Speed and Safety: Never let 'just this once' become routine. Use your checklist and escalate patterns for root cause analysis.

Communicate and Own It: Explain to your team why something's an exception or issue and what's at stake. Always close the loop.

Fostering a Culture of Transparency and Reporting

As a manager, your attitude profoundly influences whether your team embraces or resists the GRC IEM process. Encourage an environment where team members feel safe reporting potential issues or requesting exceptions without fear of blame or punishment.

Frame GRC as protection, not red tape. When people trust that management won't punish them for identifying gaps, they report issues sooner. This transparency can be an effective means of keeping organizations safer.

When an issue is identified, focus on problem-solving and root cause analysis rather than finger-pointing. When an exception is requested, engage in a dialogue about the business need, the associated risks, and the potential solutions.

When problems surface earlier, they are easier and cheaper to fix. It transforms your team into vigilant defenders of the organization, continually seeking ways to enhance processes and strengthen controls. Conversely, a culture that punishes mistakes or treats GRC as 'someone else's problem' drives issues underground. Hidden from view, they fester and grow, eventually erupting into major crises during an audit or, worse, a real-world incident. Your leadership is key to embedding GRC into your team's daily operations.

In one case, a marketing manager at a mid-sized firm wanted an unvetted social media tool. The team framed it as an exception, documented risks, and set a 60-day remediation timeline. Clear records saved them in an audit. Another team ignored a recurring DLP (Data Loss Prevention) misconfiguration, treating it as an exception. An auditor flagged the issue, resulting in $50,000 in fixes. The difference? Documentation and control.

In another example, a checklist prevented a compliance disaster. HIPAA requires specific encryption levels. A junior developer at an EMR company used a weaker encryption level, unaware of the requirement. The Director of Compliance's checklist caught the error before deployment, avoiding a critical violation and severe penalties.

A third case involved a tech lead who approved a temporary exception for database access for a critical project. Without a checklist, the "temporary" access lingered, risking a GDPR violation. A quick review caught it just in time, saving the company from a potential €20,000 (about $22,000 USD) fine.

Small Team Tip: For solo or small teams, keep it simple: Use a shared document or spreadsheet to track exceptions and issues. Even basic tools, such as the quick win checklist, can prevent confusion and support audit readiness. Hold monthly 'gut-check' reviews to validate your labels and get input from a trusted peer or manager when possible.

Embed these habits, and you'll never confuse a temporary detour with a systemic flaw again.

End-of-Chapter Checklist

- Define and document the difference between an exception and an issue for your team.

- Create or update your "Exception vs. Issue" decision checklist.

- Review two recent cases: Did you label them accurately?

- Share definitions with managers and key stakeholders to ensure alignment and consistency.

- Set a reminder to revisit and refine these definitions quarterly.

Looking Ahead

This chapter provides you with the language and tools to distinguish between exceptions and issues, setting the stage for more informed GRC decisions. In Chapter 3, we'll dive into designing and managing the exception lifecycle, using templates and workflows to keep your process tight and audit-ready.

Chapter 3: Designing and Managing the Exception Lifecycle

If You Only Remember One Thing: A Clear Lifecycle Prevents IEM Chaos

Take a moment to imagine rush hour in a city with no traffic lights or signs. There is chaos, collisions, and missed opportunities. That's what exception management looks like without structure, a straightforward setup for failure. As soon as you add clear rules, such as speed limits, stopping at red lights, and proceeding at green ones, everything flows.

In GRC, a well-designed exception lifecycle functions as your traffic system. It guides every detour safely from start to finish, keeping risk in sight and chaos at bay. Whether you're a tech startup racing to deploy code, a hospital managing HIPAA data, or a manufacturer under safety rules, this lifecycle is your shield against disaster. A clear lifecycle tames IEM chaos, turning those risky detours into smooth rides.

This chapter turns Chapter 2's theory into practice: you'll turn detours into safe journeys rather than dead ends. If you can master this, you can turn the IEM process into your ally, rather than an adversary. This lifecycle keeps traffic moving safely, with every request, every approval, and every closure under control.

The Five-Step Exception Lifecycle

Think of the IEM lifecycle as the story arc of your business. Each step mirrors what happens when an activity no longer fits a standard control. Together, these steps create the steady rhythm of exception management, the pattern you can rely on when everything else is moving quickly. Every exception follows the same five beats:

- **Intake:** The moment the challenge shows up on your doorstep.

- **Evaluation:** The point where questions deepen and the real risks come into focus.

- **Approval:** The crossroads where someone has to say yes, no, or "not like this."

- **Documentation:** The part where you capture what really happened, not what you hope happened.

- **Closure:** The ending that ties everything off so today's detour doesn't become tomorrow's problem.

Miss one of these steps, and the whole system starts to wobble. Skip one entirely, and chaos doesn't just creep in, it walks right through the front door.

1. Intake: The front door

In the intake step, you will capture requests through your GRC tool (e.g., ServiceNow, Jira, or Archer) or, for small teams, a secure shared spreadsheet. The request should capture the following:

- The policy or control being deviated from.

- Why the exception is needed (urgency, technical limitation, business need).

- The duration (longer exceptions need senior review).

- The risks and proposed safeguards.

Even small teams with limited budgets can adapt to this process. Consistency matters more than fancy tools. What matters most is that the intake process is accessible, understandable, and used every time.

Why it matters: Intake determines whether your process succeeds or fails. If it's too hard to use, you'll get shadow exceptions (untracked deviations) and surprises. If it's too loose, you'll miss crucial details.

Pro Tip: Make your intake process accessible. If it is confusing or hidden, people will work around it, and that creates bigger problems later.

A private law school once handled nearly everything through email, including maintenance, help desk, and even compliance questions. It worked fine until it didn't. When auditors asked for an audit trail, everyone scrambled through old inboxes trying to reconstruct what had happened months earlier. It was a stressful week, and accreditation was nearly derailed. That scare finally pushed the team to adopt a simple request system and stick to it.

On the other end of the spectrum, a retail operations lead under PCI DSS learned that a structured intake form didn't just keep things organized; it made vendor shortcuts impossible to hide. Once requests had to be submitted formally, questionable workarounds surfaced immediately and could be corrected before they became expensive problems.

Takeaway: Intake's your gatekeeper. Fortify it or expect shadow exceptions.

With the request in hand, it's time for investigation, the evaluation stage, where the plot thickens.

2. Evaluation: The research lab

Now, gather your experts. Each key question needs a clear owner.

- Business owners explain the "why".

- Risk or Security Assessors determine the technical or process risk.

- Compliance or Legal flag regulatory pitfalls.

Pro Tip: Gather complete information. Gaps lead to blind approval and hidden risk.

Why this step matters: Like a detective questioning witnesses, you need the whole story before deciding what to do next.

At a large software company, I watched a few promising careers stall after people approved risks they didn't fully understand. They hadn't meant to do anything reckless. They were, in fact, trying to be helpful and keep projects moving. But in security, good intentions don't protect you. That experience stuck with me. Evaluation matters because it forces you to slow down long enough to see the whole picture, including the parts you didn't think to ask about. A missing GDPR detail or unexamined dependency can turn a routine request into a costly lesson. The goal isn't to vaguely understand the exception. It's actually to understand it.

Here's a basic table you can use for assessing risk. Every organization will assess risk differently, but a simple, repeatable process is best. Remember, "KISS" now stands for Keep Information Security Simple!

Use a simple template like this to keep the risk evaluation objective.

NIST SP 800-53 controls, covering risk management, configuration management, and POA&M practices, support systematic tracking and remediation of exceptions. Many federal programs now centralize exception tracking as a best practice.

3. Approval: The turning point

This is the turning point of the case where we learn if the case is approved or rejected. Ideally, at least two approvers will share this responsibility. One would come from the security side, and the other from the business side. Either one can deny an exception request. This dual-approval approach is considered best practice in risk governance, but you should tailor it to your organization's structure and risk tolerance. Some organizations will even have a third person, often in HR or Legal to sign off as well.

- **Low-risk**: Typically, the security POC and a business manager or director.

- **High-risk**: Typically, the security POC's supervisor and an executive, generally VP (Vice President) or higher.

At one SaaS firm, we used the rule of three approvals: business lead, risk owner, and legal, ensuring no siloed decisions. However, at another technology company, we used only a security approver to approve the risk and a business approver for final sign-off.

Why this step matters: Approval is a commitment, not a rubber stamp. It often forces people into uncomfortable conversations. Most folks want to be helpful, keep projects on track, and avoid being the person who "holds things up." But sometimes saying "no" or even "not yet" is the only responsible choice.

Think of approval as a real fork in the road. Low-risk exceptions usually travel the short path — the safeguards are straightforward, the impact is minor, and the reviewers can move quickly without losing sleep. High-risk requests take a different route. They deserve a slower walk, sharper questions, and often a stop with senior leadership. That extra pause isn't red tape or needless ceremony; it's the guardrail that keeps a small oversight from turning into tomorrow's incident report or a call from a regulator you'd really rather not hear from.

Pro Tip: Always get a written sign-off and a rationale. If no one is willing to sign their name accepting the risk, stop the process.

In 2025, Apple was fined €500 million (approx. $540 million USD) under the EU Digital Markets Act (DMA) for anti-steering violations. Apple argued that its policies were necessary to protect users, but the Commission found no meaningful guardrails. The regulatory landscape is frequently changing. A potential legal challenge, often called Schrems III, may again test the EU-US Data Privacy Framework (DPF).

4. Documentation: Your defense

Record everything centrally and retain it for the required duration. **Every** part of the following list must be documented.

- Request details.

- Evaluation notes and risk score.

- All approvals and signatures.

- Mitigation steps.

- Expiration date and review schedule.

Never rely on scattered emails or chats; they vanish the moment you need evidence most. Documentation is your ongoing defense, and often the only proof you can produce when regulators, customers, or auditors come calling. If it isn't recorded centrally, it didn't happen. A central record turns SOX chaos into calm clarity for non-financial staff.

During one SOX audit, an auditor even pulled out a wizard's hat and said, "You shall not pass without documentation." It was the most lighthearted moment I've ever seen in an audit, but the message was dead serious: only solid records let you through.

5. Closure: The satisfying ending

Closure is where you find out whether the exception lifecycle actually worked. It wraps up the story arc and shows whether the safeguards you planned were carried out or quietly forgotten when the pressure faded. Set automatic reminders for review at 30, 60, or 90 days — whatever cadence works best for your environment. When an exception expires, you really only have two honest choices:

- Renew the case. Restart at Step 1, update the remediation plan, and confirm the risk still makes sense.

- Close the case. Record what happened, archive it, and move on.

Teams often skip closure because it feels administrative, especially when everyone is busy. But ignoring it leads to exception creep, the slow, subtle buildup of "temporary" workarounds that eventually become your standard operating procedure.

A simple Closure Checklist can keep you honest:

[ ] Confirm remediation
[ ] Archive records
[ ] Notify stakeholders

[ ] Log lessons learned

Here's the twist: closure isn't just the end of the process. It's one of the most reliable ways to measure your program's maturity. A team that closes consistently understands its own risks.

Practitioner's Note: Documentation Is Eternal

Documentation technically belongs to Step 4, but it's vital at every stage. By this point, most documentation is in hand and should be reviewed, but it deserves attention even during intake. Teams rarely fail because intake was insufficient or because someone missed a form. They fail because no one circled back at the end. Closure seems small, but it's where you protect your future self from digging through six-month-old emails trying to prove what was agreed upon. Treat it as preventive maintenance.

Takeaway: Without closure, exceptions never truly end; review and archive are as vital as intake.

At a fintech firm, "temporary" exceptions for unapproved tools quietly piled up. By audit time, dozens had quietly become standard practice, fragmenting data and bypassing controls–the result: a failed SOC 2 audit, proof that exception creep can unravel an entire framework.

In contrast, a healthcare client used rigorous closures to turn a string of HIPAA exceptions into process improvements, boosting efficiency and earning auditor praise, turning potential challenges into success stories.

What Not to Do: Lifecycle Pitfalls to Avoid

- Skipping intake → Leads to undocumented "shadow" exceptions.

- Rushed evaluation: Misses hidden risks.

- Verbal approvals → Evaporate under scrutiny.

- Poor documentation → Turns audits into scavenger hunts.

- No closure → 'Zombie' exceptions that haunt future audits.

In CMMC programs, unresolved gaps usually become POA&M items with defined remediation timelines. Monitor expirations carefully so temporary exceptions don't become chronic.

Bottom line: Lifecycles exist for a reason. Track your cases through each stage and set your team up for success.

Let's review the steps of the process, and who is responsible for each action:

Templates and Sample Workflows

Theory is nice, but templates make it real. Use these practical templates to build your process:

Exception Request Template (Intake Form):
- Requester name/department.

- Policy/Control deviated from.

- Reason for exception (urgency, limitation, or business need).

- Risk description (potential impact: low/medium/high).

- Proposed mitigations.

- Requested duration (including justification for longer timelines).

- Business impact if denied.

Sample Evaluation and Approval Workflow:
1. Intake → Risk team scoring.

2. Evaluation → Regular review.

3. Approval → Escalate if score >5.

4. Documentation → Central log.

5. Closure → Calendar review.

Visualize this as a flowchart: Intake → Evaluation → Approval → Documentation → Closure. (You can even draw it as a process arrow in your GRC docs!)

Practitioner's Note: Issue Requests Are Slightly Different

Mapping GRC issues to the controls they affect is a bit like a doctor connecting a patient's symptoms to the right body systems. It's how you figure out what's really going wrong. When you tie an issue (the symptom) to a control (the mechanism that's supposed to prevent the problem), the picture sharpens quickly. You can see what failed, why it failed, and whether it's part of a bigger pattern.

This kind of mapping gives the business something incredibly practical: a way to strengthen weak spots without guessing. It keeps teams from fixing the same problem twice, helps avoid redundant work across frameworks, and makes it far easier to explain what's happening to leadership in clear, non-technical terms. Instead of reacting to symptoms as they appear, you start treating the underlying conditions.

The real benefit? You stop firefighting and start building resilience. Map every issue to a control and you're not just closing tickets, you're giving your organization a clearer picture of its own health. It's the closest thing GRC has to an MRI: a scan that reveals the trouble spots before they turn into breaches or audit findings.

Quick Win: Stand Up Your First Exceptions Process in 30 Days

Ready to move from Wild West to well-oiled machine? If you're starting from scratch or fixing a messy process, use this one-month starter plan. Here's the starter plan:

- Week 1: Review your current process, identify top exceptions, and form the team.

Week 1 Checklist: [ ] Review policies, [ ] ID top exceptions, [ ] Form team.

- Week 2: Build your form, choose the tool, and test for effect.

Week 2 Checklist: [ ] Build Form, [ ] Choose tool, [ ] Test it.

- Week 3: Train your team, pilot the program, and gather stakeholder feedback.

Week 3 Checklist: [ ] Train, [ ] Pilot, [ ] Gather feedback.

- Week 4: Launch your process, monitor for gaps, and adjust your processes as needed.

Week 4 Checklist: [ ] Launch, [ ] Monitor, [ ] Adjust.

Example: In 2025, amid the EU-U.S. Data Privacy Framework (DPF) rulings, a finance team handled GDPR transfers stress-free. What was once a scramble became a streamlined, stress-free process. Use this checklist and launch in a month, like the finance team's GDPR win post-DPF.

For a manufacturing firm I consulted, we implemented this in Microsoft Teams, intake via form, evaluation in channels, and approval via an approvals app. It cut processing time by 50% and caught a safety exception (under ISO 45001, the occupational health and safety standard) before it escalated.

Tools and Resources

As you develop your exception lifecycle, consider how technology can streamline workflows, strengthen data integrity, and enhance reporting. While many dedicated GRC platforms offer comprehensive solutions, even simpler tools can be leveraged effectively. Choose tools that fit your process, not the other way around.

You don't need the latest enterprise GRC suite to manage exceptions well. The best tool is the one your team will use consistently and can scale as you grow. Here's how to match solutions, low-tech to high-tech, to each stage of the exception lifecycle.

Integrated Tools and Resources by Lifecycle Stage

1. Intake: Capturing the Exception

This is where exceptions are identified and formally reported.

- **Tool Functionality:** Centralized submission forms, automated routing based on criteria (e.g., risk level, business unit), initial data capture (description, reporter, date).

- **Examples:**

  - **Dedicated GRC** Platforms (e.g., ServiceNow GRC, Archer (formerly RSA Archer), LogicGate): These platforms typically offer robust intake modules with customizable forms, self-service portals, and automated workflows to direct submissions to the correct team for initial triage.

  - **Ticketing or ITSM** (Information Technology Service Management) systems, such as Jira, Zendesk, or ServiceNow, are often used for broader issue tracking; they can be configured with specific forms and workflows for GRC exception intake, especially in IT-centric organizations.

○ **Internal Portals** (e.g., SharePoint/Confluence): For smaller teams or those without dedicated GRC software, a simple linked form and structured list make an effective intake mechanism.

2. Evaluation: Assessing Impact and Root Cause

Once an exception is reported, it needs to be thoroughly assessed.

• **Tool Functionality:** Fields for risk assessment (likelihood, impact), linkage to existing controls/risks, root cause analysis fields, and audit trail for analyst notes.

• **Examples:**

○ **Dedicated GRC Platforms:** Allow for detailed risk scoring, mapping exceptions to your existing risk registers and control frameworks and provide collaborative workspaces for team analysis.

○ **Spreadsheet Tools (e.g., Microsoft Excel, Google Sheets):** Though less automated, structured spreadsheets work well for small teams to record evaluation details, track risk scores, and link supporting documentation.

○ **Collaboration Platforms (e.g., Microsoft Teams, Slack, Confluence):** Used for discussions, sharing evidence, and documenting insights during the evaluation process, though they would need to integrate with a more structured tracking system.

3. Approval: Authorizing the Exception

Exceptions often require review and approval from various stakeholders.

- **Tool Functionality:** Workflow automation for approvals, electronic signatures, precise status tracking, and reminder notifications for approvers.

- **Examples:**

  - **Dedicated GRC Platforms:** Fully developed platforms excel in this area, providing sophisticated workflow engines that automatically route exception requests through predefined approval chains and maintain audit trails of all decisions.

  - **Workflow Automation Tools (e.g., Microsoft Power Automate, Zapier):** Can help build custom approval workflows that integrate with other systems (e.g., sending approval requests from an intake form to an approver via email with a direct link to review).

  - **Email/Document Management Systems (e.g., Outlook, Gmail, SharePoint):** For elementary processes, approvals might be managed via email threads or documented within a central document repository, though this is less efficient and harder to audit.

4. Documentation: Creating the Audit Trail

Comprehensive documentation is crucial for audits, continuous monitoring, and future reference.

- **Tool Functionality:** Centralized repository for exception details, attached evidence (screenshots, emails, reports), version control, linkage to related policies, standards, or audit findings.

- **Examples:**

  - **Dedicated GRC Platforms:** Serve as the primary system of record, storing all exception-related data, attachments, and historical changes within a secure, auditable environment.

  - **Document Management Systems (e.g., SharePoint, Confluence, Google Drive):** Excellent for storing supporting evidence and linked documents. They should ideally integrate with your exception tracking system to ensure easy access to relevant files.

  - **Version Control Systems (e.g., GitHub for policy-as-code):** While more niche, some organizations manage policy or control documentation in version-controlled repositories, which exceptions might reference.

5. Closure: Resolving and Learning

The final stage involves verifying remediation and formally closing the exception.

- **Tool Functionality:** Fields for remediation actions, verification steps, closure notes, retesting results, trending, and reporting capabilities for lessons learned.

- **Examples:**

  - **Dedicated GRC Platforms:** Support detailed recording of remediation steps, tracking, and formal closure sign-offs. Crucially, they facilitate reporting on closure rates, outstanding exceptions, and recurring issues.

○ **Project Management/Task Tracking Tools (e.g. , Asana, Trello, Microsoft Project):** Often used to manage the specific tasks associated with remediating an exception, providing granular tracking of progress. These would feed back into the central exception system.

○ **Business Intelligence/Reporting Tools (e.g. , Tableau, Power BI, GRC platform's native dashboards):** Essential for visualizing exception trends, identifying common root causes, and reporting on the overall health of the exception management program to leadership.

Pro Tip: If you outgrow a spreadsheet, you've proven your need for a fundamental tool. That's a win, not a failure.

Other Tools and Resources

Workflow Automation/Orchestration Tools: Beyond just GRC platforms, emphasize how general automation tools (like Microsoft Power Automate, Zapier, or even custom scripts) can connect disparate systems and automate mundane tasks within the lifecycle. This empowers teams to build custom efficiencies.

Start Here: Automate intake routing and approval reminders first. These offer the fastest ROI.

Watch Out: Review automation rules often. Automation is only as good as the logic you set: unchecked rules create blind spots.

Reporting & Analytics Capabilities: Stress the importance of built-in or integrated reporting tools. For each stage, briefly mention what kind of reports or metrics are relevant (e.g., Intake: volume, categorization; Evaluation: average time to evaluate; Approval: bottlenecks; Documentation: completeness; Closure: resolution rates, recurring exceptions). This reinforces the "actionable" aspect by showing how tools provide insights.

Start Here: Begin with simple dashboards that track exception age, closure rates, and recurring risks. Even a basic spreadsheet with bar charts or a built-in reporting tool in your GRC system can give you immediate visibility and help spot trends.

Watch Out: Keep dashboards simple. Fancy visuals mean nothing if they hide key problems. Focus on clarity, accuracy, and decisions you can act on.

Integration Capabilities (APIs): Point out that the value of many modern tools lies in their ability to integrate via APIs. This enables data exchange between your GRC system, vulnerability scanners, identity management systems, and ticketing tools, creating a more holistic view.

Start Here: Connect your exception management tool to your ticketing or identity management system with basic integrations, such as syncing status updates or linking user accounts. Many tools offer out-of-the-box API connectors for the most common platforms.

Watch Out: Test and document integrations. APIs break after system updates and can expose data if permissions are too broad. Review access and log every connection.

Communication & Collaboration Tools: While not strictly GRC tools, tools like Microsoft Teams, Slack, or Confluence are vital for real-time discussion, evidence sharing, and documentation *around* exceptions. Mentioning how to use these in conjunction with structured GRC systems is very practical.

Start Here: Leverage what your team already uses: create a dedicated Slack or Teams channel for exception tracking or use shared documents for approvals and status updates. Even regular check-in meetings can boost visibility and speed up closure.

Watch Out: Capture conversations centrally. Chat and email threads disappear. Summarize critical decisions and store them in your GRC system.

With tools in place, success still depends on people: clear roles keep the system running.

Ownership and Roles: Who Does What?

No lifecycle runs on autopilot. Here's how to work well for small teams to record evaluation details, track risk scores, and link supporting documentation through the cracks. Think of roles as your cast; if they're miscast, the play flops. Assign these roles:

- Requester: Identifies the need and initiates the process.

- Risk Assessor: Usually InfoSec or GRC; assesses the request.

- Remediator: Often, the requester plans corrective actions and resolves the root cause.

- Approver: A security and a business approver.

- Documenter/Monitor: Compliance or GRC admin; tracks, closes, and archives.

- Executive Sponsor: For the highest risks, a director-level or higher leader makes the final call.

Pro Tip: Cross-train to avoid bottlenecks; they kill momentum. So does a lack of trust, which is where your soft skills come in. When people see exceptions as tools rather than punishments, you'll encourage better requests and reduce workarounds. Cross-training also builds resilience when staff turnover occurs.

In small teams, rotate roles every quarter. This spreads expertise and helps spot process gaps. With roles cast, let's see if we can supercharge them with a boost from AI.

Where AI and Automation Enhance Exception Processes

Technology isn't the villain, it's your best sidekick. Here's where modern tech can save your team hours each month, but always with a human at the wheel.

- **Intake Automation:** Use chatbots or digital assistants to guide requesters through forms and prefill common data fields.

- **AI Evaluation:** ML models scan for patterns, flag recurring requests, and match new ones to past cases.

- **Approval Workflows:** Automated routing, e-signatures, and escalation save time.

- **Documentation & Monitoring:** GRC platforms like LogicGate or Archer auto-generate reports, send closing reminders, and highlight exceptions at risk of "creeping".

- **Predictive Analytics:** AI can spot trends and flag areas where exceptions are accumulating.

- **Grok for Government:** In September 2025, the U.S . General Services Administration (GSA) launched *Grok for Government*, a pilot using the xAI API to scan patterns, flag duplicates, and suggest mitigations drawn from NIST CSF playbooks. xAI also worked with Oracle Cloud on a GDPR-compliant zero-data-retention solution. Time will tell how effective these tools have become, but the future looks bright for AI in GRC.

In one program, AI triage significantly reduced cloud exceptions, but high-risk cases were always vetted against NIST CSF. But remember: AI is a tool, not a replacement for human judgment. Always review high-risk cases in person.

AI tools can anticipate exceptions by flagging patterns aligned to NIST Cybersecurity Framework (CSF) 2.0, a widely adopted risk management model across industries.

At a midsize financial firm, exception requests used to pile up, each one waiting for a human reviewer to assign risk scores and check for duplicate submissions. When the team introduced an AI-powered triage system, everything changed.

Their AI scanned new requests, flagged high-risk ones for immediate human attention, and even suggested likely compensating controls based on historical data. Risk managers were no longer buried in paperwork; they could focus on true outliers. As one analyst put it: "Think of it as triage: low-risk exceptions move fast, while high-risk ones get escalated right away."

A global SaaS provider dreaded quarterly audits because finding all exception documentation meant hours of digging through emails and spreadsheets. That changed when they rolled out automated exception tracking with workflow integration. Now, every exception request triggers automated reminders, status updates, and evidence collection, no more scrambling for missing approvals or expiration dates. Result: audit preparation time dropped by 60%, and the team stopped dreading audit season. Their CISO summed it up: "Automation turned audit from a fire drill into a checklist."

Caution: With AI, watch for biases (recall Amazon's recruiting tool from Chapter 1) and ensure compliance with frameworks like GDPR.

In-Context, Real-World Examples and Scenarios

SaaS Startup Example: A SaaS startup hit a crunch during their SOC 2 audit when a development team submitted an exception for a block of stubborn legacy code. The intake came through a Jira ticket, and the initial reaction was predictable: tension, deadlines, and a quiet hope that the auditors wouldn't look too closely. But the lifecycle did its job. Evaluation scored the risk as high but manageable; the CTO approved it with temporary firewall restrictions; everything was documented and given an expiration date. When the code was finally replaced, the exception closed cleanly.

The audit passed without findings. The development lead, juggling releases and feature demands, found that the process turned what felt like an audit nightmare into something routine.

The Secret: The structure did the heavy lifting, not heroics.

Healthcare Example: During a system upgrade, a hospital needed temporary access to unencrypted patient data, a textbook HIPAA risk. Usually, that kind of situation sparks panic, but the lifecycle gave everyone a map. Intake captured the details; evaluation brought in the privacy officer; approval layered in monitoring; documentation happened inside the EHR platform; and closure followed once encryption was restored.

The whole thing could have spiraled into a six-figure HIPAA fine. Instead, it became a 48-hour detour that never left a mark. For the hospital administrator balancing patient care, IT pressure, and regulatory anxiety, the lifecycle turned a frightening moment into a manageable one.

Why it worked: Clear roles meant no one had to improvise under pressure.

Manufacturing Example: A manufacturing plant once bypassed safety controls to rush an order out the door, a decision that resulted in someone being hurt. Afterward, the company implemented a real exception lifecycle. Months later, when another urgent job came through, the process caught the risk immediately. Intake flagged the urgency; evaluation added temporary safeguards; approval limited the scope and duration; closure ensured the temporary measures didn't become permanent.

Since then, they've run dozens of rush orders with zero incidents. The operations manager still references that early failure as the moment the plant stopped "pushing our luck."

Key Insight: A solid lifecycle process turns rush orders into safe orders.

These examples aren't outliers; they're everyday moments in fast-moving organizations. The difference between chaos and control was never talent. It was structured. And structure only works when managers treat the process as a guardrail rather than a hurdle.

Manager's Perspective:

A strong exception process saves both time and reputation in the long run. As a manager, insist on using clear, standardized workflows for all exceptions. Don't allow backchannel approvals or context to vanish in unretained emails. Make sure your team leverages available tools (like GRC software or simple automation) to document, track, and review every step. Don't just sign off; ask questions. Like the retail case, always ask 'What's next?' to avoid hidden risks.

If AI or automation can triage low-risk exceptions or flag missing info, use it! Automation doesn't replace judgment; it frees your experts to focus on the risks that matter most.

These stories show that a lifecycle isn't bureaucracy; it's your shield in the real world. Your leadership is what ensures this process thrives. This can turn potential risks into documented, controlled business agility.

With this lifecycle locked in, you're primed for Chapter 4's battles against documentation pitfalls and audits, where we'll arm you to prevent exception creep from derailing your GRC saga.

Small Team Tip: You don't need enterprise software to run an effective process. Use free forms (Google Forms, Microsoft Forms) for intake, a spreadsheet for tracking, and calendar reminders for reviews. If you wear multiple hats, as many small-team leaders do, divide lifecycle stages among your roles or share them with a business stakeholder.

End-of-Chapter Checklist

Like story chapters, review these items to complete your IEM arc.

- Map out your current exception lifecycle, even if one person holds multiple roles.

- Identify gaps or bottlenecks at each stage.

- Select or build basic templates for intake and approval.

- Assign clear ownership for each lifecycle stage (even if it's the same person for now).

- Schedule a recurring monthly review to assess your lifecycle's effectiveness.

Now that you can manage exceptions end-to-end, the next chapter adds the final ingredient, remediation, where you transform accepted risk into measurable improvement.

Chapter 4:
Exception Requests and Remediation, Turning Risks into Rockets

If You Only Remember One Thing: Remediation Turns Exceptions into Strengths

In GRC, every risk signal is a chance to strengthen your organization or let a small crack become a crisis. Think of your car's dashboard lights: some call for immediate action (like an engine warning), while others let you keep driving long enough to get help (like a low tire pressure warning). What you never do is ignore them.

Why the Distinction Matters: Control and Intent

- **Exceptions** are controlled overrides. They are intentional, documented decisions to proceed differently.

- **Issues** are unplanned problems. Something failed, something slipped, or something broke.

A well-handled exception keeps the car on the road, while a mishandled issue can leave you stranded on the shoulder. Remediation is how you keep moving. It isn't just patching problems. It's how an organization learns, strengthens, and grows. When you look closely at why an exception is needed or why an issue surfaced, and then fix the root cause, you're not just preparing for audits. You're building a healthier, more resilient system for the people who rely on it every day.

This chapter walks you through evaluating requests, saying no with confidence when you need to, and turning every risk signal into steady forward progress.

Before we go any further, here's a quick reminder of something we covered back in Chapter 2: exceptions are **planned, temporary deviations**, and issues are **unplanned control failures**. You already know how to tell them apart; now we're going to put that distinction to work. Chapter 4 isn't about definitions. It's about what happens next: evaluating requests under pressure, communicating decisions, setting compensating controls, and turning each deviation into a tangible improvement instead of a recurring fire drill.

For a busy retail lead, spotting an issue (like an unapproved vendor tool slipping through) versus an exception (a planned detour for holiday inventory) enables quick fixes without panic. In manufacturing, an issue might be skipping safety checks on machinery without approval, risking OSHA violations. At the same time, an exception could be a planned detour for a one-time equipment test with engineer oversight. It isn't only GRC that can make this distinction, and remember, Information Security is a team sport.

Issues are breakdowns demanding investigation, while exceptions are calculated, documented deviations. Treating one as the other is a recipe for compliance chaos. Once you know whether you're handling an issue or an exception, the next step is to decide whether to approve or deny it.

When to Deny and How to Communicate It

Not every request for an exception can be granted. Denying an exception is critical to maintaining the integrity of your GRC program and protecting the organization. Most denial decisions fall into one of the following six categories.

Unacceptable Risk: Even with proposed compensating controls, the risk introduced by the exception is too high. This often involves critical data privacy, financial integrity, or core security that cannot be compromised.

- *Example:* A request to store unencrypted customer credit card data on a developer's laptop for "easier testing," even with a promise to delete it quickly. The risk of breach is too high and direct.

Better Alternatives Exist: If a compliant, secure, or less risky alternative is readily available and feasible, the exception should be denied. Always verify whether an approved or lower-risk option already meets the business goal before considering an exception.

- *Example:* A team wants an exception to use a third-party file-sharing service for sensitive documents, even though the organization already provides a secure, internal, compliant platform for that exact purpose.

Fundamental Policy Violation: Some policies are non-negotiable foundations of your organization's legal, ethical, and security posture.

Allowing an exception would fundamentally undermine these principles.

- *Example:* An exception request that would violate strict regulatory mandates like GDPR for EU citizen data, where non-compliance can carry significant administrative fines and enforcement actions by EU supervisory authorities under GDPR Article 83 (up to €20 million or 4% of global turnover).

Lack of Justification or Compensating Controls: The requester hasn't provided a compelling business reason, or the proposed compensating controls are insufficient to mitigate the risk.

- *Example:* A department requests an exception to skip mandatory security training because they're "too busy," with no proposal for how they'll acquire the necessary knowledge or mitigate the risk posed by unaware employees.

Lack of a Robust and Credible POA&M: The requester (or action plan owner) hasn't provided a plan of action and milestones that inspire confidence, either because the plan isn't realistic, isn't resourced, or doesn't include precise dates, owners, and interim controls.

- *Example:* A team asks for an exception to keep using a legacy system that doesn't support encryption. Their proposed POA&M says they will "look into options this year" and "upgrade when feasible." The plan has no dates, no project approval, no interim safeguards, and no real commitment. With no clear path to remediation and no compensating controls, the exception is denied. A POA&M must show how the risk will be reduced and when. If it cannot show that, the safest answer is no.

Convenience as the Sole Reason: While convenience is a factor, it rarely outweighs security, compliance, or significant risk.

- *Example:* A developer wants to bypass standard code review processes just because it's "faster" for a minor bug fix.

Key Takeaway: Denials foster innovation and trust when communicated well.

Good compensating controls are effective and measurable

Strong IEM programs are not built on distrust of teams, but on professional realism. Most control gaps are not caused by bad intent. Instead, they arise from ambitious delivery timelines, aggressive sprint goals, and limited capacity to absorb unplanned work. Effective GRC simply acknowledges this reality.

By consistently questioning whether proposed controls are verifiable and sustainable, even early-stage programs can improve resource planning, reduce operational friction, and materially increase the likelihood that remediation efforts succeed.

Since not all compensating controls are equal, let's look at what makes one effective versus ineffective. Good compensating controls are specific, enforceable, measurable, and genuinely reduce risk. They inspire confidence in a way that inadequate compensating controls do not.

1. **Data minimization and de-identification:**

   ○ **Control:** Using only *essential* data, stripped of any direct identifiers (e.g., anonymized names or numerical codes instead of actual employee IDs), in the unapproved service. All original, fully identified data will remain on our secure internal systems.

   ○ **Why it's good:** It directly reduces the impact of a potential breach by making any exposed data less sensitive or unusable on its own. It's a clear, actionable technical step.

2. Enhanced monitoring with alerting:

- **Control:** Put real-time monitoring in place for traffic to and from the unapproved service, with clear alert thresholds. When something looks unusual or crosses those thresholds, the Security Operations Center is notified immediately so they can investigate.

- **Why it's good:** Provides early detection capabilities if a problem occurs. It's measurable (alerts generated, response time tracked) and doesn't rely solely on human vigilance.

3. Strict access controls with multi-factor authentication (MFA):

- **Control:** Access to the unapproved service will be limited to only the three essential project members, who must use institutionally provisioned laptops and strong Multi-Factor Authentication (MFA) for every login, regardless of the service's default settings.

- **Why it's good:** Significantly reduces the risk of unauthorized access to the service, even if credentials are stolen. It's enforceable and strengthens user authentication.

4. Short, non-renewable closure clause with forced data deletion:

- ○ **Control:** The exception is valid only for a strict 30-day period. After that, all data for the unapproved service will be automatically purged, and the account will be deactivated. Be sure to document these actions to maintain a clear audit trail.

- ○ **Why it's good:** Limits the exposure duration, prevents "exception creep," and ensures a definite end to the risk. It's specific and verifiable.

Inadequate compensating controls are vague, unverifiable, or ineffective

Inadequate compensating controls are often generic, rely solely on good intentions, are challenging to verify, or don't actually mitigate the specific risk introduced by the exception. They offer little to no absolute assurance.

1. We'll be careful / We'll try harder:

- ○ **Control:** The team promises to exercise extreme caution when using the unapproved service.

- ○ **Why it isn't good:** This is not a control at all. It's a promise, not a safeguard. It isn't measurable or enforceable and depends solely on consistent human follow-through. It provides zero actual risk reduction.

2. Increased awareness:

- ○ **Control:** The team will ensure all team members are aware of the risks associated with using this unapproved service.

- ○ **Why it isn't good:** Awareness is essential, but it doesn't *prevent* or *detect* security incidents stemming from system vulnerabilities or misconfigurations. It's a general training concept, not a specific control against a concrete technical deviation.

3. Manual daily checks of logs:

- **Control:** A team member will manually review the service's access logs once a day for any suspicious activity.

- **Why it isn't good:** While better than no review at all, manual log checks are inefficient and difficult to sustain consistently; they rely on individual follow-through rather than automated enforcement, lack clear criteria for what constitutes "suspicious" activity, and are easy to deprioritize during routine operational demands. As a result, they do not scale well and provide weak, inconsistent detection.

4. It's only for a short time:

- **Control:** The risk is minimal because the team will only use the unapproved service for 2 weeks.

- **Why it isn't good:** Time itself is not a control. A data breach or security incident can unfold in minutes or hours. Relying on duration alone ignores that the vulnerability exists for the entire period. Duration does not mitigate the risk itself; it only limits the window during which it may occur.

I've seen teams offer "We'll be careful" as a control in a biotech firm, only to have it backfire during a HIPAA audit, turning a minor exception into a significant headache.

Practitioner's Note: Denials Spark Innovation

A denial isn't a dead end; it's a detour to better paths. Frame it as collaboration: *Let's find a compliant way to hit your goals.* Many colleagues rarely interact with the IEM team, so each encounter is an opportunity to build trust and effective working relationships. Remember that denials don't have to be the end of the conversation; often, they're the start of collaboration. Look for other solutions.

Use this checklist during review meetings to quickly evaluate the strength of proposed controls.

Compensating Controls Checklist

1. Is it specific and measurable? (Daily exports vs. *We'll check sometimes*)

2. Does it reduce real risk? (MFA vs. *Be careful*)

3. Is it enforceable and verifiable? (Automated alerts vs. manual logs)

4. Does it limit exposure? (30-day cap vs. open-ended)

When reviewing exception requests, managers should challenge vague controls like *we'll be careful* and insist on measurable safeguards. This is where you, as a leader, ensure significant risk reduction, not just paper compliance. When an exception is denied or closed, use the discussion as a springboard for innovation rather than a source of frustration.

How to Communicate a Denial

Denying an exception can be delicate, as the requester often believes their need is legitimate. Effective communication is key to maintaining trust and cooperation.

- **Be clear and concise.** State the denial unequivocally.

- **Explain why.** Don't just say "no." Explain the specific risks, policy violations, or lack of justification that led to the denial. Reference relevant GRC policies or regulatory requirements.

- **Offer alternatives and next steps.** Provide a path forward. Provide compliant options or suggest a pathway to achieve their objective within policy boundaries. For example, a tool may become viable once the required security integrations are in place.

- **Maintain professionalism and empathy.** Understand the requester's business needs and communicate with respect. Frame each denial as a protective measure for the organization and its customers, and not as a personal judgment.

- **Document everything.** Record the request, the justification for denial, the communication, and any alternatives offered. This is crucial for audit trails.

Cautionary Tale: The Rubber Stamp That Sunk an Audit

A Cloud service provider was facing a persistent problem: multiple development teams were requesting exceptions to use unapproved open-source libraries in their production code. Official policy stressed that only vetted libraries could be used for client-facing software. The developers, under tight deadlines, kept submitting the exception requests.

Under pressure from the developers and their leadership, who were accountable for the deadlines, GRC approvers began rubber-stamping these exceptions. They were approved with only superficial reviews and compensating controls, such as *increased code review,* that were vague and unenforceable. There was a refusal to pursue root cause analysis and to push back, allowing GRC to serve as a *rubber stamp* out of fear of "impeding the business".

Then came the SOC 2 audit. A sprawl of unapproved libraries, some containing known vulnerabilities that standard scans had missed, lit up the auditor's results. The vulnerabilities traced directly to GRC's pattern of rubber-stamping exception requests instead of enforcing controls. The auditors cited this as evidence of a systemic breakdown in change management and vendor risk processes, leading to a qualified SOC 2 opinion with multiple control exceptions under the AICPA SOC 2 criteria for Security and Availability.

Client trust eroded quickly, and one key customer canceled a multimillion-dollar contract over the perceived lapse. While approving the exception may appear to support business velocity in the short term, it often has the opposite effect. Rubber-stamping requests without enforceable controls or remediation paths leaves risks unresolved, creates downstream rework, and slows progress toward real remediation. Over time, this approach reduces both security posture and operational efficiency rather than improving either.

Often, communication is both the problem and the solution. At a university, a professor wanted to use a new cloud-based analytics platform and proposed several compensating controls, such as anonymizing data where possible, limiting access to her team, and storing analysis results internally. The exception had to be denied due to the strict data-handling policy; fortunately, it was communicated clearly. The reasons were clear: there was no direct contract with the university, no audit rights, and no explicit data-protection assurances.

GRC and the professor collaborated on a solution using a vendor product already within the university's ecosystem. The proactive approach, driven by effective communication regarding a denied exception request, paved the way for a better solution that benefited the professor's work and was safe for the university and its students.

Every new GRC exception request is an opportunity for the team to ask whether there is a process, tool, or training fix that could prevent this from happening in the future. The answer should be logged; these responses should be reviewed monthly. Proactively working through this list of questions and answers can result in fewer requests, as GRC can anticipate the trends based on this extension of communicating as a partner.

Failures like this underline why clear communication and cultural buy-in matter just as much as technical accuracy.

Practitioner's Note: Overcoming Business Pushback

Teams may resist remediation by pointing to time, cost, or perceived bureaucracy. Effective GRC leaders anticipate these objections and prepare to address them with clear business impact. By using cost comparisons, relevant case studies, and practical context, you can meet stakeholders where they are and more effectively gain their buy-in.

Acknowledge operational disruption while consistently reframing remediation as both risk reduction and business enablement. Successful discussions typically lead with facts and context, supported by empathy and clear financial realities. Over time, positioning remediation as an investment in organizational resilience, not an administrative hurdle, builds durable stakeholder support.

What Not to Do: Remediation Pitfalls

Even experienced teams can fall into these traps. Be vigilant about remediation.

- Don't rubber-stamp renewals without review; it's exception creep's best friend.

- Avoid "fix it later" promises without deadlines; they breed debt.

- Never deny without alternatives; it erodes trust.

- Don't skip RCA. You're inviting repeats.

- Ignore metrics? You'll fly blind into audits.

Bottom line: Remediation is key. An IEM process without remediation is like a sales program without revenue. But what if a denial sparks pushback? Let's talk escalation.

Communication and Escalation Framework

Even the best exception programs need a straightforward process for escalation and appeal when disagreements arise. Sometimes, a denied exception request isn't the end of the conversation. For significant requests or those with high business impact, a formal escalation and appeals process is vital. This ensures fairness, transparency, and critical decisions are made at the appropriate organizational level.

Why Escalation is Needed:

- **High Business Impact.** The denied exception may significantly impede a project or revenue stream.

- **Complex Risk Profile.** The risks involved are nuanced and require a broader perspective or specialized expertise (legal counsel or other specialists).

- **Disagreement on Risk Assessment.** The requester genuinely believes the GRC team has misjudged the risk or the effectiveness of their proposed compensating controls.

- **Policy Ambiguity.** The GRC policy itself might be unclear or not fully applicable to the unique situation.

For complex profiles, escalation often involves input from legal or other subject-matter experts, particularly for regulatory nuances such as GDPR.

The Escalation Process:

Typically, a formal escalation process moves the decision-making authority up the chain.

1. **Initial Review.** The GRC team reviews and makes an initial decision.

2. *Managerial Escalation.* If denied, the requester's manager might escalate to a higher-level GRC manager or a departmental head.

3. **Risk Committee/Steering Committee.** For high-impact or contested denials, the request might go before a dedicated Risk Committee or a cross-functional Steering Committee that includes senior leadership, legal, IT, and business unit

heads. This committee has the authority to weigh the comprehensive business benefits against the aggregated risks.

4. **Board Level (Rare).** In extremely rare cases, usually involving enterprise-level risk or significant strategic implications, the appeal might even reach the Board of Directors or its relevant subcommittee.

The Appeals Process

An appeal is a formal mechanism for a requester to challenge a denial. It's not about ignoring rules but ensuring due process and a comprehensive review. The appeal typically requires three things:

- A written justification for why the initial denial should be overturned.

- Any new information or revised compensating controls.

- A clear statement of the business impact if the exception remains denied.

A robust escalation and appeals process ensures that challenging GRC decisions are made transparently and receive appropriate senior-level buy-in. It provides a safety net for legitimate business needs without compromising the integrity of the GRC framework.

Let's see how escalation can work well in practice.

A company in the entertainment industry sought to bring a better product to market. An executive earmarked a cloud-based rendering and collaboration platform that wasn't on the company's approved vendor list. The Visual Effects (VFX) supervisor admired the accelerated rendering speed and real-time collaboration features. Those features, however, ran directly afoul of Intellectual Property (IP) rules regarding pre-release content on unvetted third-party platforms.

When the exception was denied, the VFX supervisor escalated to the Head of Production. Security and Legal again rejected the request, as the risk of a pre-release leak of high-value IP remained too significant. Despite the denial, the team implemented a sandboxed environment that isolated the collaboration and rendering tools from external exposure, enabling the use of the desired features while protecting pre-release IP from leakage.

Turning Exceptions into Improvements and Insights

This is where GRC evolves from a cost center to become a strategic value driver. Every exception request, whether granted or denied, and every detected issue is a valuable data point. They are symptoms of something more profound: a policy gap, a systemic vulnerability, a lack of training, or a technology limitation.

At a retail chain, unreviewed exceptions slowly expanded the PCI DSS scope and weakened controls within the Cardholder Data Environment (CDE). What began as a single holiday-season bypass went unchecked and gradually spread across the supply chain, resulting in fragmented security and a near-breach that could have led to significant fines. By implementing a quarterly review cycle to prune expired exceptions and reset control boundaries, the team stabilized the environment and avoided further incidents.

The data points of your case inventory can lead to insights into your processes and drive improvements through Root Cause Analysis (RCA), dedicated process improvement, and strategic insights.

Root Cause Analysis: The Why Game. For every exception or issue, the most critical step is to keep asking *why?* This practice is known as Root Cause Analysis.

- **For an Exception:** Why was this deviation *needed*? Was the policy too rigid? Was the standard tool insufficient? Did a new business need emerge that wasn't foreseen?

- ○ *Example:* If multiple teams are repeatedly requesting exceptions to use a specific type of cloud analytics tool that isn't on your approved vendor list, the root cause may be that current approved tools lack a critical feature the business genuinely needs.

- **For an Issue:** Why did the control fail? Was it human error (lack of training, oversight, or supervision)? A system bug? A process breakdown (no review step)?

- ○ *Example:* If unencrypted data was found on a server (an issue), the root cause might be a developer bypassing security checks due to a rushed deadline, and a lack of automated enforcement in the deployment pipeline.

Process Improvement: Fixing the How. Once you understand the root cause, you can implement targeted improvements.

- **Policy adjustments:** If a policy is consistently proving impractical or creating too many legitimate exceptions, it might need to be refined, updated, or made more flexible where appropriate.

- **Technology investments:** If a system limitation is a recurring cause of exceptions, it highlights a need for new technology or upgrades.

- **Training and awareness:** If human error is the root cause of issues or exceptions, it points to a need for better training, more straightforward guidelines, or more frequent awareness campaigns.

- **Automated controls:** If manual processes are error-prone and lead to issues, look for ways to automate those controls (e.g., automated checks in a deployment pipeline).

Strategic Insights: Informing What and Where. Aggregating and analyzing data from exceptions and issues provides powerful strategic insights for leadership:

- **Identifying systemic weaknesses:** A pattern of exceptions related to a particular type of data handling or a specific department might indicate a broader systemic weakness in your control environment or a particular risk domain.

- **Informing resource allocation:** High volumes of exceptions or recurring issues in a specific area (e.g., cybersecurity, data privacy) can justify increased investment in tools, staff, or training for that area.

- **Shaping future strategy:** Insights can inform business strategy, risk appetite, and investment decisions. For example, if a legacy system is constantly generating high-risk exceptions, it strengthens the business case for a complete modernization project.

- **Culture of accountability:** Transparent exception and issue management fosters a culture in which employees understand that GRC is a shared responsibility, not just a set of rules imposed by a separate department. In retail, this might mean allocating resources to fix PCI DSS gaps in supply chains; in SaaS, it could justify NIST CSF training to slash repeat issues.

By learning from every deviation, organizations turn reactive problem-solving into a proactive engine for continuous improvement and strategic advantage.

Metrics, KPIs, and Continuous Improvement

Once lessons are identified, measurement ensures they lead to progress. *What gets measured gets managed.* This adage is profoundly true in GRC. To safeguard your exception and remediation processes, ensure they are effective and continuously improving by tracking key metrics and Key Performance Indicators (KPIs). Metrics are your GRC GPS. They keep you on track to your destination.

Key Metrics and KPIs to Track:

- **Number of open exceptions:** This gives you a baseline of your *deviations*. Track it by type (e.g., security, privacy, operational), by department, and by business owner.

- **Average age of open exceptions:** How long are exceptions staying open? Longer ages indicate stalled remediation, unaddressed technical debt, or a lack of accountability.

- **Exception closure rate:** The percentage of exceptions resolved within their approved timeframe or by their sunset date.

- **Number of exceptions converted to permanent solutions:** This is a crucial indicator of GRC maturity; it shows whether issues are truly resolved or merely extended. Are you just extending exceptions, or are you truly fixing the root cause?

- **Number of repeat issues:** How often do the same compliance issues or control failures occur? High numbers here indicate failed remediation or incomplete root cause analysis.

- **Remediation completion rates:** The percentage of assigned remediation tasks completed on time versus delayed.

- **Time to resolve issues:** How quickly are new issues being addressed from identification to closure?

Why Measure Matters

- **Track progress:** See if your GRC program is improving over time.

- **Identify bottlenecks:** Pinpoint where approvals get stuck, where remediation is lagging, or where specific teams are struggling.

- **Demonstrate value:** Show leadership and auditors that your GRC efforts are practical and contribute to risk reduction.

- **Drive accountability:** Assigning metrics to owners encourages responsibility.

- **Support resource allocation:** Use data to justify tools, staff, or training for the areas with the most significant risks.

- **Root cause analysis effectiveness:** Track how many issues or exceptions include a documented cause and follow-up plan.

Practitioner's Note: RCA Isn't Optional

Root cause analysis isn't a *nice-to-have;* it's your detective kit for GRC mysteries. Skip it, and you're treating the symptoms, not the disease. Use the *Five Whys* technique. In this method, you ask *why* five times to drill past symptoms, like uncovering a training flaw behind recurring NIST CSF vulnerabilities.

Regularly reviewing these metrics in GRC dashboards and reports enables organizations to shift from a reactive stance to a proactive, data-driven, continuous-improvement approach. It's like getting regular health check-ups for your GRC program; you identify problems early and adjust your "lifestyle" accordingly.

If you see the average age of open exceptions climbing past 90 days, it's time to review for process gaps or bottlenecks. Use the information and see how it changes regularly. Track against SOC 2 Trust Services Criteria, demonstrating operational control over customer data, or ISO/IEC 27001 requirements to spot systemic gaps early, like fine-tuning an engine before it overheats.

While frameworks like SOC 2 and ISO 27001 are technically voluntary, many industries treat them as contractual prerequisites. In practice, their influence is effectively mandatory for organizations that want to compete in regulated or enterprise markets.

Picture your dashboards as a GRC command center that turns raw metrics into a story of progress and alerts. Copy it into your system and tweak the numbers for your team, whether tracking HIPAA-training completion, BAA (Business Associate Agreement) renewals, or PCI DSS fixes, and watch accountability soar.

Visualizing Success: The GRC Metrics Dashboard

Imagine steering your GRC ship through choppy waters, exceptions popping up like rogue waves, issues lurking like hidden reefs.

In the IEM process, indicators are your radar. The leading ones spot storms on the horizon, while lagging ones tally the damage after the fact.

Leading indicators are proactive whispers that predict risks before they swell into crises, like a finance team's early flag on GDPR data gaps to avoid enforcement actions.

Lagging indicators are the rearview mirror, confirming what has already gone wrong. For example, a retail chain might track PCI DSS non-conformities or cardholder data incidents within its Cardholder Data Environment (CDE) after an audit or security event.

I have laid out some information on indicators in the table below for quick scanning, with cross-industry examples.

Vanity metrics in KPIs are like polishing the deck while your ship springs leaks; they dazzle on dashboards but sink real progress. Take "exceptions closed without RCA": It looks impressive, say a 95% closure rate flashing green for a manufacturing exec under ISO 27001 pressure, but without digging into root causes, you're just putting Band-Aids on fractures.

Vanity metrics risk allowing recurring issues to fester, breeding exception creep that erodes trust, invites adverse audit findings or qualified attestation results, and waste resources on endless firefights.

In one case from my InfoSec work, a SaaS team focused on closing issues quickly rather than resolving underlying root causes. That approach eventually contributed to a failed SOC 2 assessment when unaddressed patterns culminated in a security incident, reinforcing that effective KPIs must drive meaningful remediation, not just surface-level progress.

Quick Tip: Start your dashboard with free tools like Google Sheets: track open exceptions by department, add charts for trends, and review quarterly. Tie it to frameworks: Benchmark exception ages against your organization's internal remediation SLAs aligned to SOC 2 Trust Services Criteria and ISO 27001 control objectives to spot gaps early. Chapter 6 revisits dashboards and provides a link to a very crude beginning of one you can use to start.

Takeaway: Metrics that drive improvement, not just appearances, are the heart of continuous GRC progress. But measuring is only part of the equation. Let's look at how to act on those insights to prioritize and tackle what matters most.

Once you understand what your metrics reveal, the next step is deciding which risks to address first.

How to Prioritize Remediation

Measurement shows you where the risks lie, and prioritization ensures you fix the right ones first. Not every issue or exception carries the same weight. When you have a backlog of remediation tasks, how do you decide what to tackle first? That's why prioritization is key to effective resource allocation and risk management.

Prioritization Frameworks

1. **Risk-Based Prioritization**: The gold standard. Focus remediation on issues or exceptions that pose the highest organizational risk.

   ○ **Impact:** How severe would the consequence be if the risk materialized or policy deviation caused a problem? (e.g., massive financial loss, major data breach, significant reputational damage, regulatory fines, or loss of life).

   ○ **Likelihood/Urgency:** How likely is it to happen, or how quickly does it need to be addressed? (e.g., an active security vulnerability vs. a theoretical future risk).

- **Approach**: Address the highest-impact, highest-likeli-
  hood items first. A data breach vulnerability should take
  precedence over a minor formatting issue on an internal
  document.

In a healthcare clinic I advised, risk-based triage turned a backlog of
HIPAA compliance gaps into a practical roadmap for modernization.
In one SaaS case, root-cause analysis of recurring cloud access issues
exposed policy gaps, leading to a complete overhaul that prevented
breaches.

1. **Regulatory urgency:** Some remediations aren't optional;
   they're legally mandated.

   - **Approach**: If a regulatory body (such as the SEC, an EU
     Data Protection Authority under GDPR, or the HHS
     OCR under HIPAA) issues a finding or deadline, it
     jumps to the top of the list. Non-compliance here carries
     severe legal and financial penalties.

2. **Cost-Benefit analysis:** Sometimes, it's about optimizing
   your investment.

   - **Approach:** Compare the cost of remediation against
     the potential cost of inaction (e.g., potential fine, lost
     business, brand damage). Prioritize fixes that offer the
     greatest risk reduction at the lowest cost.

3. **Resource availability and feasibility:** Be realistic about
   what your team can accomplish.

   - **Approach:** While high-risk items are paramount, some-
     times a lower-risk item might be a "quick win" that can

be resolved immediately with available resources, freeing up time for bigger tasks. Don't let the perfect be the enemy of the good. Steady, practical progress is what matters.

If GDPR penalties loom, data privacy exceptions go straight to the top of your triage list. While, in general, it is preferable to be proactive rather than reactive, avoiding penalties and negative consequences is quite essential!

Tips for Prioritization

- **Clear ownership:** Every remediation task needs a clearly accountable owner.

- **Defined deadlines:** Assign realistic but firm deadlines.

- **Regular review:** Review the remediation backlog with stakeholders and leadership to adjust priorities as risks evolve or resources change.

- **Break down big tasks:** Large remediation projects (like a system overhaul due to technical debt) should be broken into smaller, manageable phases with individual deadlines.

Quick Win: Remediation Triage Matrix

For day-to-day prioritization, a simple triage matrix is highly effective. It's a two-by-two grid to help quickly categorize and decide on action.

Using this matrix helps teams make quick, consistent decisions about where to focus their efforts, ensuring the most critical risks are addressed first, while still keeping an eye on important but less urgent improvements.

Pilot this in a team meeting, using your backlog as a live exercise. Tackle high-urgency, high-impact items first, for example, a PCI DSS vulnerability under Requirement 11 (vulnerability management testing) should take priority over minor UI fixes.

A growing retailer reduced open exceptions by nearly 60% in just two quarters by tightening triage discipline and holding teams accountable for closing what they opened. They didn't get there by adding friction. They got there by treating exception handling as part of building a resilient, execution-driven organization, not as a barrier between teams.

For Managers: Handling Exception Requests and Remediation

How you handle exception requests sets the tone for your entire risk culture. Don't rubber-stamp requests, even when you trust the team making them. Ask for clear rationale, practical compensating controls, and a realistic plan to fix the underlying root cause.

If you need to deny a request, explain the reasoning and work with the team to find a safe alternative. A denial shouldn't feel like a door closing. It should feel like a chance to improve the process or clarify expectations.

Track meaningful metrics: How many open exceptions are active? How long have they been sitting? Are the same issues reappearing? And when a request could materially affect the business, escalate early to your risk or steering committee.

Every exception is a signal. Treat each one as an opportunity to learn something, strengthen something, or prevent a repeat. Then share those lessons across the business so everyone moves forward together.

Managing GRC exceptions and issues isn't about rigid adherence to rules for their own sake. It's about intelligent risk management, continuous learning, and strategic improvement. By distinguishing exceptions from issues, communicating effectively, prioritizing wisely, and pursuing remediation with discipline, IEM transforms challenges into catalysts for growth, resilience, and sustained success.

Remember: every request, denial, and remediation affects real people. This can be project teams under pressure, IT leads trying to meet deadlines, and even end customers trusting your organization with their data. The best GRC programs pair technical excellence with empathy and clear communication, making compliance a partnership, rather than a punishment.

When managers set the right tone and track measurable outcomes, this is what success looks like in practice.

What Winning at IEM Looks Like

Picture a tech firm where exceptions spark upgrades, like turning legacy headaches into modern wins. If the risk remains too high or better alternatives exist, the request is denied, but the denial is communicated not as a "no," but with clear explanations of *why* (e.g., unacceptable data exposure risk) and of the available *compliant alternatives*, fostering trust rather than frustration.

For complex cases or those with significant business impact, the firm has a clear escalation and appeals process. Denied requests can be elevated to a cross-functional Risk Committee, comprising senior leaders from legal, IT, and business units. This ensures that high-stakes decisions are reviewed comprehensively, weighing all perspectives while upholding the integrity of the GRC framework. This structured approach prevents "rubber-stamping" and ensures accountability at the appropriate level.

Crucially, every exception request and issue is viewed as a learning opportunity. Winning IEM teams meticulously track metrics and KPIs, such as the number of open exceptions, average time to remediation, and repeat issues, often revealing systemic patterns. For example, a recurring pattern of exceptions for legacy system integrations might trigger a strategic initiative to modernize that infrastructure, turning a recurring headache into a long-term solution. This data-driven approach allows them to identify root causes and prioritize remediation effectively, using a risk-based triage matrix that focuses resources on high-impact, high-urgency items (e.g., critical security vulnerabilities) over minor compliance observations.

This continuous feedback loop, from initial evaluation and careful decision-making to data analysis and strategic improvements, transforms the IEM team from a reactive compliance burden into a proactive system for building resilience, optimizing operations, and maintaining a competitive edge in a dynamic regulatory landscape.

Small Team Tip: Prioritize ruthlessly. Tackle high-risk exceptions first, and automate approvals for low-risk or recurring scenarios. Use a simple triage matrix to decide what *must* get your attention. Share remediation wins with leadership; visibility helps justify more help as you grow.

End-of-Chapter Checklist

- Review how exception requests are currently evaluated.

- Document your process.

- Update your criteria for approval and denial.

- Implement a remediation triage matrix or priority list.

- Track remediation progress for all open exceptions.

- Share at least one remediation success story with your team or leadership.

Next: Chapter 5 shifts from fixing issues to proving and sustaining those fixes through effective documentation, audit preparation, and control discipline.

Chapter 5: Documentation, Audit, and Preventing Exception Creep

If You Only Remember One Thing: Documentation Is Your Audit Lifeline

An audit is your chance to show your work. Strong documentation is the evidence that your controls are operating as intended and your risks are being managed responsibly. When records are clear and current, audits stop being a source of stress and become a way to confirm real progress.

In GRC, skimping on documentation doesn't just invite findings; it can unravel your entire operation. Done right, however, it's your secret weapon for resilience. Building on Chapter 3's lifecycle, this chapter dives deeper. We'll craft audit-proof records, dismantle common pitfalls like permanent exceptions, and equip you to stomp out exception creep before it poisons your processes.

Whether you're a retail exec dodging PCI DSS traps or a manufacturing lead navigating ISO 45001 audits or a defense contractor pursing CMMC certification, you'll emerge audit-ready and confident.

Our goal here in Chapter 5 is to turn potential gaps into airtight records. In Chapter 4, we focused on resolving exceptions effectively. Now we explore how to prove that success with a documentation process that tells a clear story from the opening of the request to its resolution.

What to Document, and How

Documentation isn't busy work. It's the backbone of your GRC story. Your documentation proves every exception was a calculated detour, not a reckless swerve. What should your documentation capture? Everything that tells the tale:

- **Origin:** Who, what, and why was the request made?

- **Evaluation:** Risk scores, stakeholder comments, supporting evidence.

- **Approvals:** Who signed off or rejected the case, with their reasoning.

- **Mitigations:** Safeguards such as temporary firewalls, enhanced monitoring, or limited access.

- **Closure:** Proof the exception was resolved or renewed, with dates, signatures, and a summary.

Keep documentation practical and easy to maintain. Whether you use a GRC tool like Archer or a secure shared drive, the goal is to keep everything in one place rather than bury it in email threads. Make sure records include timestamps, version history, and clear approvals so changes are easy to follow. Simple habits such as labeling files clearly, attaching screenshots or logs, and writing short summaries of what happened and why help create documentation that stands up to audit review.

As a manager, your role is to ensure your teams *adhere* to these documentation standards. This isn't just about compliance; it's also about building a collective memory for your organization that speeds onboarding and preserves lessons learned. Documentation is your armor. If you miss details, fines can mount quickly.

Practitioner's Note: Don't Wait Until the Audit to Check Your Docs

The best GRC programs don't treat documentation as a once-a-year scramble. Set quarterly "mini-audits" for your own team. You'll spot gaps while they're small, keep exceptions fresh, and walk into real audits with confidence. I've seen teams go from dread to calm just by practicing. When questions came, the answers were ready.

Remember: Routine "mini-audits" turn audit season from a scramble into a victory lap.

During my time in InfoSec at a SaaS firm, we documented a cloud-migration exception thoroughly and attached the evaluation discussions and supporting rationale directly to the case. When our ISO 27001 audit came around, that documentation passed review without issue and saved the company nearly $50,000 in rework. Just as importantly, it preserved the reasoning behind our decision so future teams could build on it instead of starting over.

Pro Tip: This narrative format is also a big winner when very high-risk cases are brought to C-suite executives (e.g., CIOs or CFOs) for approval. This level of approver is often further removed from the exact issue that the person assessing the risk and proposing the remediation plan. Proper documentation, especially in a concise, narrative style, can make a tremendous impact in assisting the executive to understand the context of the situation.

Technical View: If you're the engineer or admin handling these exceptions, remember: your documentation isn't just *for compliance*, it's protection for you, too. Good records make root cause analysis and remediation easier and protect you if questions arise later.

With these elements captured, let's look at tools that make audits a breeze.

Checklists, Forms, and Passing Audits

Audits aren't ambushes; they're predictable tests. But only if you're prepared. Start with checklists to ensure nothing slips. A basic audit prep list might include:

[ ] Set expiration dates for all exceptions

[ ] Link documentation to relevant policies (e.g., HIPAA for healthcare)

[ ] Practice responding to auditor questions ("Why this mitigation?")

[ ] Scan for signs of exception creep (e.g., renewals over three times)

[ ] Confirm compensating controls for legacy risks (e.g., network isolation under HIPAA)

[ ] Review for AI-related exceptions per the EU AI Act ties to GDPR.

Forms amplify this: Create a standardized exception log form with fields for lifecycle stages, pulling from Chapter 3's template but adding audit hooks (fields that let auditors quickly trace evidence) such as "Regulatory Impact (e.g., PCI DSS compliance?)" or "Evidence Attached (Y/N)."

Passing audits boils down to transparency, and if you can show auditors the whole arc, from intake to closure, and tie it to frameworks like SOC 2, you set yourself up for success.

At a biotech startup, the firm used a simple form to document a temporary bypass of encryption during a system upgrade. In a 2025 biotech push for HITRUST compliance, the form's audit trail meant no significant findings.

An excellent tip for non-GRC folks: Practice *audit theater*, a rehearsal exercise where teams role-play auditors' questions. Conduct these role-play reviews quarterly to spot gaps, making the real thing feel like a dress rehearsal, not opening night jitters. This role-play meant the biotech's HITRUST audit was a breeze.

How to Run Audit Theater

1. Gather your team for a 30-minute session.

2. Assign one person as the "auditor" to ask tough questions (e.g., "Why was this exception approved?").

3. Use your one-page audit trail to answer, practicing clear, concise responses.

4. Note gaps (e.g., missing documentation) and fix them before the real audit.

5. Repeat quarterly to build confidence.

In early 2025, a mid-sized European bank faced heavy scrutiny under GDPR. During a system merger, a manager flagged a temporary exception for sharing customer data. Without documentation, it could've spiraled into a breach and penalties.

Instead, they used a standardized exception log with fields for risks, approvals, safeguards, and regulatory tie-ins. The form spelled out the "why" (merger timeline), the safeguards (encrypted channels, limited access), and the expiration (30 days). When auditors probed, the clear trail, complete with approvals, proved it was a controlled detour. No fines, no findings. Just relief and proof that even non-experts can pass audits with a one-page form.

Practitioner's Note: Auditors Aren't the Enemy

Especially if you're new to audit cycles, remember that auditors are your coaches, not your opponents. Their job isn't to catch you, but to spotlight weak points before they become big problems. Treat them as guides, not gatekeepers.

Remember: The best audit outcomes happen when you treat your auditor as a resource for improvement.

Exception Creep, Time Limits, Review Cycles

Exception creep occurs when temporary fixes linger long past their intended lifespan. It typically shows up when exceptions are renewed without confirming they are still needed or have been properly mitigated. This allows a one-off exception to turn into a routine risk. As discussed in Chapter 3, weak or inconsistent case closures often start the cycle.

The most effective way to prevent exception creep is to apply firm time limits and maintain regular review cycles. In most environments, exceptions should expire within 90-180 days unless formally reassessed. Quarterly reviews that ask basic questions, such as whether the exception is still needed and whether the underlying risk has been mitigated, help keep programs disciplined. Time limits are not always easy to enforce but overlooking them almost always increases operational and audit risk.

How to Build a Review Cycle:

1. Set automated alerts at 30/60/90-day intervals.

2. Involve cross-teams for fresh eyes. For example, a hospital used this to prune HIPAA exceptions, cutting open cases by 30%.

In a finance firm I advised, creep undermined GDPR compliance as 'one-off' data-sharing exceptions multiplied unchecked, leading to a near-miss breach until quarterly cycles pruned them back.

In retail, I've seen PCI DSS compliance creep. A chain's repeated vendor exceptions led to fragmented security, culminating in a breach. Counter it with automation and accountability: automated reminders paired with KPI-driven reviews (for example, 'reduce open exceptions by 20% quarterly'). This isn't micromanaging; it's hygiene that keeps your operation lean and audit-resilient.

I've seen this in manufacturing. In one plant, legacy equipment triggered ISO 27001 exceptions that piled into technical debt, nearly causing a shutdown until reviews cleared them. In another factory, a 'permanent' exception allowed non-compliant machinery to meet deadlines while ignoring ISO 45001 safety rules. In both cases, deferred remediation compounded future costs and risks.

This debt and unaddressed wear led to a breakdown, costing downtime and fines. The lesson? These pitfalls aren't one-offs: they're chains that link back to poor hygiene, eroding trust, and inviting auditors to dig deeper. Spot them early by auditing your exceptions quarterly, asking: "Is this still necessary, or just a habit?"

Common Pitfalls: Permanent Exceptions, Technical Debt, and Legacy Risks

Some pitfalls are like when a sports team makes unforced errors. Allowing permanent exceptions to linger, technical debt to accrue interest, and legacy risks to remain unpremeditated can haunt a team. Legacy risk is especially pronounced in more technology-centric companies, where legacy risks can manifest as unpatched vulnerabilities that invite exploitation under scrutiny.

Permanent exceptions often create the illusion of temporary relief while quietly weakening systems over time. Industry research has shown that technical debt now consumes a significant share of IT budgets, diverting resources that could otherwise be used to improve security and resilience. When exceptions allow low-priority fixes to linger indefinitely, organizations end up paying the "interest" through increased operational risk and constant workarounds.

Over time, these legacy risks compound. Outdated systems that no longer align with modern security standards can turn minor gaps into serious exposure and elevate the likelihood of a breach.

Legacy Risk Example: A healthcare provider used a 25-year-old patient record system that was no longer supported. Upgrading would cost millions and take years, so they requested a temporary GRC exception to keep it running safely through network isolation and other compensating controls.

The case was approved with those compensating controls in place. Auditors reviewed these under SOC 2's "continuous monitoring" criterion, so logging was critical.

While these measures contained the risk, the core problem remained: an irreducible legacy system that required long-term exception management until modernization was possible. Under HIPAA, such systems often fail to meet ePHI (electronic Protected Health Information) encryption requirements; isolation buys time, but actual fixes require migration.

Spotlight: The Myth and Risk of Permanent Exceptions

Permanent exceptions are a GRC myth. The reality is that "permanent exceptions" are red flags for process immaturity, inviting audit hits and disguising risk. They mask root causes, breed complacency, and then explode during incidents (e.g., a "permanent" vendor access gap that leads to a data leak). Bust the myth! Set hard time limits, require reapprovals, and treat them as issues (per Chapter 2).

As one tech client learned the hard way, "permanent exception" often means "permanently vulnerable."

Case in Point: A Permanent Exception That Came Back to Haunt

Early in my career, I consulted for a healthcare provider racing to integrate a new EMR system. Leadership approved a 'permanent' exception for unencrypted legacy data transfers, citing budget constraints and promising a future fix. That decision went unexamined for two years until a HIPAA audit uncovered it as a significant exposure.

The root cause was a lack of periodic review. The exception became routine, quietly accumulating technical debt with every renewal. The issue was resolved by enforcing 90-day limits and escalating long-running exceptions to senior leadership. The key takeaway is simple: permanent exceptions don't fade away; they remain active risks until they are formally closed.

Technical Debt Example: A colleague described a fast-growing startup that rushed its trading platform to market. Development shortcuts left behind an undocumented system that intermittently generated errors, forcing teams to step in and clean things up manually.

Since the policy aligned with PCI DSS and required fully automated reconciliation, the company filed a GRC exception. It was granted with compensating controls, including daily manual reviews by senior analysts and additional logging, but the underlying technical debt remained.

The exception was renewed repeatedly, auditors flagged it as non-conformity, and the workaround consumed resources while introducing human error. True fixes would have demanded code refactoring, but scans and controls only bought time and increased technical debt.

What Not to Do: Avoid These Documentation Disasters

Table 5.1. Common documentation disasters

Avoid these common traps. Each one has sunk an otherwise solid program.

Bottom line: Keep everything documented, with unique details for each case, time-stamped, centralized, and regularly reviewed. These missteps have turned small control gaps into six-figure audit findings. Avoid them, and you'll never be the cautionary tale.

Exception management isn't just about controls or paperwork. It's about people. When teams trust that exceptions will be managed fairly, reported openly, and never used as a 'gotcha' weapon, they'll flag risks early and own their part of the process. Good GRC culture encourages transparency, not perfection.

Escalation Paths and Exception Hygiene

When exception creep looms or risks spike, escalation paths are your signal flares, alerting leadership before conditions ignite. Define clear escalation tiers: low-risk cases route to managers, medium-risk to directors, and high-risk to executives or committees. Exception hygiene means ongoing maintenance. Winning teams prune expired cases, audit for patterns, and foster a 'no-shame' culture where issues are flagged early.

In 2025, a bank's escalation path helped avoid GDPR-scale penalties, proving that hygiene really is a lifesaver. Technically, run scans to catch debt. In compliance, align exceptions with regulations. For management, frame hygiene as ROI through reduced rework.

Quick Win: One-Page Audit Trail Template

Audit trails don't need novels; keep it to one page for quick wins. Here's a plug-and-play template:

If you only implement one new habit after this chapter, make it this: Use a one-page audit trail for every exception, and keep it up to date. It's the fastest way to go from audit panic to audit poise. Print this, fill it out for each exception, and store it centrally. Even better, you're your own tailored version. Review for exception creep quarterly and start with one team as a pilot.

Pro Tip: Test this template on one exception and track how much review time it saves.

Takeaway: Every 'temporary' exception is a potential future audit finding. Regular review and strong documentation are your only real defenses. Pilot with one exception case to slash review time.

Manager's Perspective:

Documentation isn't just paperwork. It is what protects the organization during an audit and provides clarity when something goes wrong. As a leader, champion thorough, narrative-style documentation for every exception. Don't wait until audit season to request records. Instead, schedule brief monthly or quarterly "mini-audits" within your team. Encourage transparency over perfection.

A documented mistake is easier to address than a missing record. Most importantly, review open exceptions regularly for signs of creep, including items that remain open too long or are repeatedly renewed without resolution. Your involvement sets expectations and pushes strong exception hygiene to become part of daily operations.

Small Team Tip: Automate what you can, even in small ways like setting calendar reminders for exception reviews and expirations. Standardize documentation using a copy/paste template. In 2025, free tools can still handle GDPR creep alerts seamlessly.

If you're both 'doer' and 'checker,' ask another department to spot-check your process once a quarter. This reduces blind spots for tiny teams.

End-of-Chapter Checklist

Check these items off for audit preparation.

- Create or update your exception documentation template.

- Audit a sample of recent exceptions. Are they fully documented?

- Identify any "permanent" exceptions. Schedule them for review.

- Set calendar reminders for periodic exception reviews and expirations.

- Establish an escalation path for unresolved or high-risk exceptions.

- Check for framework updates (e.g., SOC 2 exceptions).

Mastering documentation and review cycles means you aren't just surviving audits; you're transforming your GRC into a proactive source of resilience and organizational memory. In Chapter 6, we take a deep dive into reporting, dashboards, and creating a culture of accountability.

Chapter 6: Reporting, Dashboards, and Creating a Culture of Accountability

If You Only Remember One Thing: Transparency Builds Trust

P icture a sales team trying to navigate around GRC in a fog of uncertainty, wary of the so-called "department of no." Transparent dashboards flip that script: like lighthouses in the haze, they reveal fair processes and build trust. In one finance firm I consulted, opaque reporting fueled shadow IT chaos. When reports became clear and accessible, collaboration followed, cutting repeat exceptions by 30%. This chapter builds on Chapter 5's remediation, integrating reporting and dashboards into your navigational toolkit to foster a culture of accountability.

Why Reporting Matters More Than Ever

In most organizations, *reporting* evokes groans: bloated slide decks for leaders and quarterly reviews that feel disconnected for staff. Executives can't steer blind; teams can't work in the dark; and auditors require clarity.

Reporting bridges strategy and execution by replacing noise with clarity. It answers three essential questions: Where are we now? How do we measure up? What's next?

For a retail manager juggling PCI DSS compliance or a hospital administrator navigating HIPAA, clear reports transform GRC from "extra work" into a decision-making edge. They turn regulatory obligations into practical tools that guide daily priorities.

After fixing issues, we need a way to measure whether our fixes hold. That's where reporting and dashboards enter the picture.

Dashboards, Metrics, and ROI

Dashboards act as your GRC lighthouse, distilling complex data into an at-a-glance narrative. But their value isn't in visual polish, it's in alignment with fundamental business objectives. Without that connection, charts are decoration, not decision support. The following section clarifies what dashboards are actually meant to accomplish.

We live in a world of data, but value only emerges when information is organized toward action. Metrics define what we measure; dashboards translate those measurements into insight. Time and effort are the inputs, clarity and decisions are the returns. What matters most is connecting that insight back to business objectives and ROI: knowing not only what the numbers say, but why they matter.

What Makes a Good Metric?

A metric is only as good as its clarity and relevance. Use the SMART framework, which is often used for creating goals. SMART stands for

Specific, Measurable, Achievable, Relevant, and Time-bound. This is a simple framework that keeps goals (and our metrics) focused and actionable. Just as the exception lifecycle guided remediation, the SMART framework guides measurement.

Actionable GRC Metrics

- **Exception Aging:** Percentage of exceptions older than 90 days, segmented by severity.

- **Control Effectiveness:** Percentage of high-risk controls failing internal tests during the current quarter.

- **Training Completion:** Percentage of staff completing required training on time.

- **Incident Trends:** Year-over-year change in reported phishing incidents per 1,000 employees.

Weak Metrics

- **Number of Dashboards Created:** Tracks activity rather than outcome.

- **Security Policy Awareness:** Too vague to drive specific action or improvement.

- **Compliance Audit Completed:** Binary status lacks depth; what matters is the number, severity, and remediation of findings.

Practitioner's Note: Start with what Leadership Already Values

If uptime is the executive priority, show the risk's impact on uptime. If cost is the priority, show exception-related cost avoidance. GRC gains traction faster when it speaks the organization's native language. There are many ways to fail, but giving leadership what they value isn't one of them!

Now that we've distinguished good from bad metrics, the following examples illustrate what strong measurement looks like in practice:

These measurements become meaningful when they connect directly to business outcomes.

Demonstrating ROI

Every GRC leader eventually hears the same question: *What value do we actually deliver?* Having a clear, defensible answer matters, not for theory, but for credibility when it counts. Take a cybersecurity exception program: if dashboards show a 30% reduction in repeat exceptions because issues are remediated earlier, that's measurable ROI. When visual reporting cuts audit preparation from six weeks to two, the impact is obvious. Even softer wins, like fewer late-stage escalations, signal value when translated into business terms.

In a SaaS startup I advised, the GRC lead used dashboards to highlight how streamlined exception tracking, enabled through targeted NIST CSF adjustments, reduced repeat requests by 25%, freeing developers for a new product feature launch. The team didn't just avoid delays, they accelerated delivery, earning executive recognition and an all-hands shout-out. Proving ROI isn't just about survival; it fuels innovation, confidence, and forward momentum.

Think of ROI as a map: cost avoidance shows where value is protected, efficiency gains trace the path forward, and business enablement marks the destination.

Stories like this matter even more amid today's economic headwinds. The "Great Resignation" of the early 2020s quickly gave way to what many now call the "Great Layoff." U.S. tech-sector job cuts totaled roughly 93,000 in 2022, surged past 200,000 in 2023, and exceeded 150,000 again in 2024 (per Layoffs.fyi and TrueUp). So far in 2025, cumulative layoffs have already crossed 160,000. In this environment, departments must demonstrate operational value early and often. Cost avoidance may never make headlines, but as cases like Wells Fargo's $3 billion remediation cost prove, it represents one of the most powerful—and often overlooked—forms of return on investment.

What is GRC's ROI?

The returns of a strong IEM program are harder to show on a spreadsheet than, say, a product sold in stores. The value of the program is evident in several ways.

Cost Avoidance: Regulatory fines, legal settlements, customer remediation costs, and enforcement actions are among the largest avoidable losses organizations face. Wells Fargo alone has paid more than $3 billion in penalties related to compliance failures, while Meta, Binance, Credit Suisse, and others have incurred fines exceeding $1 billion.

While the exact dollar value of "avoided fines" can't be modeled with precision, the economic reality is straightforward. Every major compliance failure prevented represents millions (or even billions) of dollars preserved. Cost avoidance is not hypothetical ROI; it is one of the most direct and material financial returns delivered by effective GRC programs.

Efficiency Gains: Streamlined GRC processes reduce manual effort, speed up approvals, and free up resources for strategic initiatives. This allows the team to be proactive, focusing on higher-value activities like root-cause analysis to prevent recurring issues.

Business Enablement: Through well-defined exceptions and clear risk communication, GRC helps the business innovate faster without unnecessary risk. When risks are accurately documented and regulatory requirements understood, the organization can pursue opportunities with confidence.

Reputation Protection: The intangible, yet critical, value of maintaining customer trust and brand image by preventing security incidents or compliance failures cannot be easily discounted.

Reporting, metrics, and dashboards show transparency and value, but they should also connect to strategy. Metrics aren't just numbers; they should tell a story that informs strategic decisions and resource allocation.

Pro Tip: Well-known GRC tools (Archer, ServiceNow GRC, LogicGate, etc.) can automate metric collection, while smaller teams might leverage spreadsheets with pivot tables for basic tracking. Where automation is feasible, it should be explored.

Visuals That Drive Management Action

Managers are pressed for time and need immediate, clear information. Well-designed dashboards translate complex data into insight. Three principles separate actionable reports from those that get ignored.

1. **Clarity Over Complexity:** Use red/yellow/green indica-

tors sparingly for meaningful thresholds. Remember, Keep Information Security Simple (KISS)!

2. **Context Over Raw Data:** "8 exceptions this month" means nothing without context; is it high, low, or normal?

3. **Trend Over Snapshot:** Trends turn a static snapshot into a story. As an example, rising exception ages might signal an audit storm ahead.

4. **Audience Matters:** Tailor dashboards for CISOs (risk details), team leaders (action items), or boards (high-level trends).

5. **Tell a Story:** Flow from overall risk posture to specific problem areas.

A Practical Example

How does this look in practice? Once we understand these reporting principles, we can turn metrics into a compelling story for leadership.

Example: From Data Dump to Story

Before: "We had 57 open exceptions in Q2."

After: "Open exceptions decreased by 20% from Q1 to Q2 due to faster triage of low-risk requests and targeted training for high-risk business units. If this trend continues, we will reach our target of under 30 open exceptions by year-end."

The second example turns numbers into narrative, which drives decision-making.

Types of GRC Visuals and Their Use Cases:

• **Trend Charts:** Show progress (e.g., closure rate over time).

- **Heatmaps:** Highlight risk exposure (likelihood vs. impact).

- **Pie/Bar Charts:** Break down exception types or show the top violations.

- **Status Indicators:** Red/yellow/green for control health.

- **Top N Lists:** Top 5 recurring issues or departments.

Case Study: Transforming Audit Dread into Dashboard Delight.

A retail chain once struggled with PCI DSS compliance, relying on static spreadsheets that produced long monthly readouts but little clarity. The underlying data existed, but it wasn't actionable. The GRC team reorganized those same metrics into an Excel dashboard highlighting open exceptions by type, average age, and control effectiveness (red/yellow/green). Managers immediately spotted hotspots, cut audit preparation time by 50%, and uncovered a supply chain vulnerability that saved thousands in potential fines.

The same principle applied at a manufacturing plant. Dashboards visualizing ISO 45001 metrics flagged safety gaps before a near-miss incident, reducing preparation time and averting fines. In both cases, translating raw data into clear visuals moved leaders from passive reporting to proactive action.

What Not to Do: Dashboard Overload

- Don't track everything: select metrics that drive action.

- Avoid cluttered charts or dense text; simplicity tells the story.

- Never use metrics to "gotcha" teams; partner, don't punish.

Manager's Perspective:

As a manager, don't just consume dashboards; *demand* dashboards that answer your strategic questions. Work with your GRC team to refine them until they are truly actionable decision-making tools, not just data dumps. Ensure your reports and dashboards meet leadership's requirements and provide answers to their strategic questions.

Once you're measuring what matters, the next question becomes: how do you prove it matters? That's where ROI enters.

Building Trust Between GRC Teams and Business Units

Let's acknowledge the elephant in the room: a traditional divide exists between many business units and even IT teams, who often view GRC as an obstacle—or worse, the so-called department of no. Effective reporting and metrics can become a two-way street that fosters not just compliance but collaboration.

Dashboards and reports alone can't repair fractured trust, but they can earn it when paired with consistent behavior. Many GRC teams unintentionally position themselves as gatekeepers rather than partners. That dynamic is within GRC's control to change. When teams are only seen during moments of enforcement, mistrust grows; when they engage earlier as problem-solvers, collaboration follows.

Risk management shifts from automatic refusal to protecting business value and the reputations of fellow teams. Transparency must move in both directions, sharing not just decisions, but the reasoning behind them, so partners see GRC not as a roadblock, but as a guide.

Key Strategies to Build Trust

- **Co-create Metrics:** Involve business units in choosing metrics; ownership fosters support.

- **Share Early and Often:** Regular updates prevent surprises and allow teams to course-correct.

- **Explain the "Why":** Frame risks in business terms (e.g., "This could cost $5M in GDPR fines") and solve together.

- **Use Empathy:** Understand business pressures and constraints so that risk recommendations reflect real-world trade-offs.

- **Align Goals:** Link GRC metrics to business KPIs (e.g., uptime, revenue).

- **Avoid jargon**: Telling a stakeholder, "Policy 3.1.2 nonconformance," can alienate, but opting for something more like "We risk losing $1M if this fails" can connect.

- **Avoid data hoarding.** Keeping risk data hidden is the breeding ground for suspicion.

The Quarterly Review That Changed Everything

At a financial services firm, GRC clashed with IT over mounting exception requests that were often rubber-stamped. A new dashboard revealed 78 overdue exceptions. Instead of launching another corrective review, the GRC team invited IT to co-design a shared dashboard using three metrics: time to remediate, repeat requests, and training completion.

Co-creating metrics shifted the dynamic from being "measured against" to being "measured with." That same collaborative approach had succeeded earlier at a technology startup, where jointly owned compliance reporting cut SOC 2 exception volume by half. Within two quarters, exceptions at the financial firm dropped by 40%, and IT began proposing new training initiatives, turning friction into partnership.

Transparency in Decision-Making

Clearly communicating the reasons for decisions, especially denials, is essential. When business units understand the policy basis and risk rationale behind a decision, defensiveness gives way to dialogue. Just as important is creating real feedback channels and being prepared to treat that feedback not as resistance, but as insight. This is even more important when it arrives bluntly or imperfectly.

At a large software company, a divisional security officer and our IEM team initially operated at cross-purposes. The tension was never personal, but perspectives were misaligned, and communication was inconsistent. To reset the relationship, we began short, regular check-ins and, critically, acted on the feedback we received. Over time, this transparency and follow-through rebuilt trust.

His feedback was candid, frequent, and often sharper than most teams are comfortable hearing. But he was right: feedback is a gift. By listening without defensiveness, incorporating what mattered, and collaborating openly, the relationship shifted from friction to partnership.

When budget pressures later threatened staffing reductions, he became one of our strongest advocates with senior leadership, demonstrating how fully the trust had turned. What some initially labeled a "problem stakeholder" proved to be one of our most valuable allies once communication became honest and collaborative.

Case in Point: *Biotech Startup & GRC Team Find a Win-Win.*

Scenario: When I worked for a biotech startup, we had a specialized medical practice that used our SaaS product integrated with sophisticated equipment. Because of labor shortages and a desire to save money, the team wanted to use WiFi to send patient data. Our compliance team initially denied the request due to concerns about HIPAA violations.

Trust-Building in Action: Instead of a hard "no," my supervisor asked me to engage the vendor in dialogue, understanding their needs and communicating ours. Within a couple of meetings, we collaboratively found an alternative within the medical provider's approved ecosystem that met both physical topology and security requirements.

Outcome: The facility got its solution up and running securely, we maintained compliance, and, perhaps most importantly, built a stronger relationship. The takeaway here is that trust is earned through transparency and partnership, not perfection. Co-created dashboards become shared scorecards, not weapons. Finally, don't forget that feedback is a gift, and should be treated as such.

Reducing Exception Requests Through Education and Awareness

Back in Chapter 3, we went through the exception lifecycle. We learned how to manage deviations once they occur. The next step is more ambitious: reducing the need for exceptions in the first place.

That requires a two-pronged approach. We educate our users on when exceptions are truly appropriate and what information helps resolve cases efficiently. At the same time, we educate ourselves through lessons learned and root-cause analysis so we can eliminate recurring exceptions before they ever need to be filed.

Education for Case Reduction

Exception volume rises when policies are poorly communicated, when teams misunderstand intent, or when outdated guidance creates unnecessary friction. Reducing requests starts with education, both of the business and of GRC itself.

Policy Walkthroughs: Host short, plain-language sessions that explain not just what policies require, but why they exist. Business units usually interact with only a small set of policies; walk through those together until expectations are clear. Internally, GRC must hold itself to the same standard by reviewing policies annually for vagueness or obsolescence. Ambiguous or outdated policies lead to avoidable exceptions.

Role-Based Training: Replace generic training with role-specific guidance for developers, project managers, executives, and team leaders, so content connects directly to daily workflows. Within GRC, cross-train the team to avoid single points of failure and knowledge bottlenecks.

Accessible Resources: Create living resources: searchable wikis, FAQs, and simplified policy summaries. These tools reduce submission errors and rework for staff, while preventing information silos inside the GRC team itself. Shared visibility prevents cases from slipping through the cracks.

Highlight Success Stories: Demonstrate how early engagement with GRC avoids rework and accelerates approvals. Celebrate when root-cause analysis or improved documentation eliminates repeat issues. Recognizing both user engagement and internal process improvement builds lasting momentum.

When teams understand the why behind policies, case volume naturally declines, and when GRC continuously learns from exceptions, it becomes less reactive and more proactive. This only works when metrics aren't treated as static reports but as tools for continual learning and improvement. In practice, this two-way education creates a feedback loop: users feel empowered, GRC stays sharp, and exception requests begin to taper off.

In a hospital environment, root-cause analysis of HIPAA exceptions exposed training gaps. Targeted workshops closed those gaps and significantly reduced repeat requests.

Practitioner's Note: Trends Are Your Friends

Trend exceptions, by department. If one team submits 40 percent of all requests, it's probably a communication issue, not a people problem. Address the root cause rather than approving endless exceptions.

Education Example: A finance firm sees a recurring pattern of GDPR-related exceptions for data transfers to third parties.

Intervention: The GRC team analyzed root causes and identified a gap in understanding about compliant data transfer mechanisms. They launched targeted workshops, updated the "Data Transfer Playbook," and posted the update on their wiki.

Result: A significant reduction in GDPR-related exception requests, improved compliance, and increased confidence in handling sensitive data.

Recognizing and Rewarding Positive Behavior

Accountability isn't only about identifying failures; it's equally about celebrating wins. Positive reinforcement builds momentum and reinforces desired behavior.

Examples:

- Publicly recognize teams that close exceptions ahead of schedule.

- Feature "exception of the month" where a creative workaround eliminated a risk.

- Incorporate GRC performance into performance reviews (carefully balanced with fairness).

- GRC Champions: If your team identifies and empowers an advocate in other business units, it's a win for everyone involved.

Picture knighting a developer as 'GRC Champion' for early vulnerability flags. This helps to slash repeats, like rewarding a retail team's PCI DSS hygiene to inspire enterprise-wide wins. These feedback loops form the foundation of the continuous improvement we'll explore next.

What Not to Do: Performative Recognition

- Don't make recognition feel performative or political; it must be authentic.

- Don't overemphasize negative outliers; celebrating wins should be the goal, not finger-pointing.

Recognition creates allies. Business units that feel seen for good behavior are more likely to partner proactively on future initiatives. Be seen as the department of "go", not the department of "no".

Manager's Perspective:

As a manager, you have an incredible opportunity to shape behavior. Actively seek out and celebrate instances where your team demonstrates strong GRC hygiene. Recognition makes GRC a valued part of daily operations.

Proactive Tactics: Office Hours, Weekly Syncs, and More

Dashboards are a feedback loop; proactive engagement keeps the loop healthy. Too often, GRC is reactive, responding to exceptions, incidents, and audits rather than preventing them. Simple habits shift this dynamic:

- **Office Hours:** Hold regular, informal drop-in sessions where business units can ask questions, discuss potential issues, or seek advice before submitting requests formally. This catches problems early.

- **Weekly Syncs:** A 15-minute stand-up between GRC and high-risk business units can surface issues before they escalate.

- **Pilot Feedback Groups:** Test dashboards with a small audience, gather input, and refine before wide rollout.

- **Embed GRC Into Business Units:** Assigning a GRC staff member to liaise with a business unit can offer insights otherwise impossible to attain.

- **"Lunch and Learn" Sessions:** Informal educational sessions on specific GRC topics relevant to various departments.

- **Pre-Mortems/Risk Workshops:** Proactively engage with new projects or initiatives to identify and address GRC risks *before* they become problems.

Office hours are similar to those of preventative medicine. Hosting office hours can catch 'symptoms' before they become full-blown issues.

Takeaway: The best dashboards fail without human follow-through. Pair reporting with regular dialogue to turn insights into improvement.

Case Study: Manufacturing Firm's Proactive Safety Shift.

Before: Safety-related exceptions under ISO 45001 were handled reactively, often only after near-miss incidents. GRC was viewed as a compliance gatekeeper rather than a safety partner.

After: The GRC team launched weekly safety syncs with production managers, established monthly GRC office hours directly on the factory floor, and integrated GRC advisors into early project planning for new equipment and facility changes.

Result: Safety exception volume dropped steadily over subsequent quarters, employee participation in hazard identification increased, and reporting shifted from after-the-fact documentation to early risk escalation, marking a clear transition from reactive compliance to a proactive safety culture.

Quick Win: Dashboard-in-a-Day

When time or resources are scarce, start simple. A Dashboard-in-a-Day exercise delivers visible impact fast. Zig Ziglar once said, "By the mile, it's a trial. By the inch, it's a cinch." You don't have to overhaul the GRC IEM program in one day; build your dashboard one metric at a time. Next, we have a quick win that will enable us to create the dashboard, one metric at a time.

Let's understand the scope: We're starting today with what we have. We are not waiting for budgetary approval for a fancy GRC tool that we don't have. If we have one, great; if not, open up a spreadsheet. We're going to focus on a Minimum Viable Product (MVP) that provides us with clear insight into our Issues and Exceptions environment.

Steps:

1. **Identify 3-5 Key Metrics:** Focus on the most impactful ones (e.g., Open Exceptions, Average Age, Closure Rate, Top 3 Recurring Issues).

2. **Gather Data:** Pull data from your current exception/issue tracking system (even if it's just a spreadsheet).

3. **Choose a Tool:** Excel or Google Sheets are perfect for this quick win.

4. **Create Simple Visuals:**

   a. Bar chart for "Open Exceptions by Type."

   b. Line graph for "Average Age Trend."

   c. Simple table with R/Y/G conditional formatting for "Closure Rate."

5. **Add Context:** Include a small text box for "Key Insights" or "Call to Action."

6. **Schedule a Review:** Share it with a pilot group (e.g., your direct manager or one business unit leader) and get feedback.

You can download a starter version of this dashboard at gruntwork s.tech/resources. It is deliberately lightweight, designed as a functional foundation that demonstrates structure and flow without requiring advanced Excel skills. The template is not a finished product, but a working example meant to be customized to your organization's data, workflows, and reporting needs. Use it as both a learning tool and a launchpad for building your own GRC metrics program.

At one SaaS firm I advised, these early dashboards sparked a NIST CSF pre-mortem that identified cloud risks before launch, helping the team proactively mitigate issues.

Practitioner's Note: Perfection Vs. Good

Perfection kills momentum. A basic dashboard delivered quickly can spark leadership buy-in, unlocking resources for deeper reporting later.

Explain your chart's methodology. If using conditional formatting, ensure a casual observer can easily grasp its meaning.

Pro Tip: Don't wait for the perfect tool. As a manager, empower your team to build a quick, simple dashboard. The act of visualizing even basic GRC data will immediately spark conversations and reveal insights that scattered data never could.

When your first iteration has gotten approval from your manager or a knowledgeable peer, it's time to extend the pilot to a business unit. Focus on trends and make the dashboard useful to the recipient team. This means it has to be tailored to some degree. CISOs, the sales team, developers, and legal all need very different information.

Take the feedback you receive from your pilot group(s) and incorporate it. Then prepare for broader distribution.

Case Study: Turning Data into a Conversation

A healthcare organization faced 250 open exception requests, some of them years old. A new CISO introduced dashboards that segmented aging metrics by business unit and used the data to frame the backlog as a shared problem rather than a blame exercise. That shift sparked a three-month sprint that closed 60 percent of open cases.

Case Study: Celebrating the Quiet Wins

In a global manufacturing company, phishing simulations showed improvement in some plants but stagnation in others. The dashboard spotlighted a plant in Ohio that improved click rates from 18% to 2% in one quarter. The GRC team publicly recognized the plant manager, not just the numbers. Other plants soon mirrored Ohio's practices, driving enterprise-wide improvement.

Case Study: The CFO's "Aha" Moment

At a law school, a skeptical CFO dismissed dashboards as "noise" until a simple bar chart revealed a trend: spikes in unresolved exceptions aligned with costly audit findings. Suddenly, GRC's "noise" became a financial story worth telling. The CFO became an advocate overnight.

Creating a Culture of Accountability

Dashboards and metrics are tools; the culture they create is the real outcome, one where teams own risk without blame and leaders act before problems escalate.

- Teams own risks without fear of punishment for discovering and reporting issues.

- Leaders prioritize remediation proactively, rather than reactively.

- GRC is embedded in daily operations, not treated as an afterthought.

Four Levers for Culture Change

Like a compass guiding through risks, these levers turn dashboards into cultural anchors, transparency as your edge.

1. **Model Behavior at the Top:** Executives who review dashboards in leadership meetings set the tone.

2. **Integrate with Business Goals:** Link risk metrics to revenue, customer satisfaction, or time-to-market.

3. **Communicate Progress Relentlessly:** Celebrate improvements, don't just flag failures.

4. **Simplify the Ask:** The easier dashboards are to access and interpret, the more likely they will be used.

Transparency as a Competitive Advantage

In high-performing organizations, GRC dashboards are not dusty compliance artifacts; they are strategic instruments. They surface blind spots, highlight progress, and build trust. When transparency becomes a habit, not a quarterly exercise, GRC evolves from gatekeeper to guide.

Remember: Transparency builds trust; trust drives accountability; accountability fuels improvement.

Now that we are using data effectively through reporting and dashboards to create a culture of accountability, it's time to bring it home in Chapter 7, where we learn about the risks of the next generation and the importance of continuous improvement.

Small Team Tip: Don't overcomplicate dashboards; one clear chart (exceptions by risk, age, or owner) in Excel or Google Sheets is enough. Schedule a monthly "show and tell" with leadership to demonstrate impact. Transparency is your best tool for scaling influence.

End-of-Chapter Checklist

To reinforce these ideas, use the following checklist to turn reporting concepts into daily practice.

- Identify key exception metrics to track (e.g., open exceptions by risk, by age).

- Build a simple dashboard or report (Excel, Google Sheets, or your GRC tool).

- Schedule a regular cadence for sharing exception metrics with stakeholders.

- Recognize and reward at least one example of positive exception management behavior.

- Gather feedback on your reports and adjust them for clarity and usefulness.

Chapter 6 showed how dashboards and reporting create transparency, build trust, and embed accountability into daily operations. But risk never stands still. In Chapter 7, we'll look ahead at emerging risks, next-generation challenges, and how continuous improvement keeps GRC relevant and resilient.

Chapter 7: The Path Forward: Next Generation Risks and Continuous Improvement

If You Only Remember One Thing: Adapt or fall behind.

GRC programs that stand still quickly become irrelevant. Emerging technologies, evolving regulations, and shifting business priorities demand that exception management keep pace. The organizations that thrive are those that anticipate change, adjust rapidly, and embed improvement into their culture, not just their dashboards.

GRC is an evolving ecosystem: new species (technologies, regulations) appear, old ones fade, and the balance of power shifts. Teams that adapt thrive; those that don't get left behind. In sports, they say, "the best ability is availability." **In GRC, the best ability is adapt-**

ability. That constant evolution doesn't just reshape cybersecurity; it changes how every GRC exception must be identified, approved, and tracked.

Why it Matters: In today's accelerated world, new technologies (AI, quantum computing, IoT), evolving regulations (new privacy laws and industry-specific standards), and sophisticated threat actors emerge daily. A static GRC practice will quickly become irrelevant, leading to increased risk exposure, audit failures, and ultimately, business disruption.

A small, innovative software company was preparing to launch a new AI-driven analytics product when it learned a highly restrictive, industry-specific data-privacy regulation would take effect in six months. These changes would pull the rug out from under a core feature of their product.

Initial reactions were understandably tense. The development team saw a complete re-architecture, legal saw insurmountable roadblocks, but their GRC lead didn't think the sky was falling. He formed a cross-functional team and adopted a "lean compliance" approach. They met regularly, broke significant changes into small, actionable sprints, and built data handling prototypes. Within five months, the team not only met the new mandate but also engineered a more robust, privacy-centric product that became a selling point. Their success wasn't luck. It was adaptability applied through disciplined exception management.

Adaptability isn't just about survival; it's about seizing unforeseen opportunities. Lifecycle management, documentation, and reporting are your foundation, but GRC excellence depends on how you build and adapt from there. It's how leading organizations evolve their exception management practices.

That same adaptability also extends to newer areas like ESG (Environmental, Social, and Governance). Whether or not leaders view it as central to their mission, ESG is now a practical risk concern. Investors increasingly expect it as a marker of stability. Regulators, including the SEC, mandate climate disclosures. Markets often punish companies that lag behind their peers. In practice, ESG is less about ideology and more about resilience and long-term stability.

For GRC teams, the takeaway is simple: ESG can create exceptions just as cyber, privacy, or financial issues can. Supply chain disruptions, reporting gaps, or governance lapses may trigger the same scrutiny as a failed control. CEOs already rank ESG among the top risk factors for 2025. These expectations are formalized through frameworks such as TCFD and the EU's CSRD, which now require measurable ESG risk disclosures. Treating ESG as another layer of resilience helps organizations anticipate volatility and keep stakeholder trust.

While ESG reframes *what* we must manage, technology is transforming *how* we manage it. To stay ahead, GRC must evolve together with modern tools.

Technology has become the bridge between policy intent and operational execution. The same adaptability that governs ESG must also drive how we use AI, automation, and cloud. The next wave of tools, from AI to automation, is redefining the pace and precision of exception management.

How Modern Tech (AI, Automation, Cloud) Shapes Exception Management

The tools and risks shaping GRC today are fundamentally different from those we managed even five years ago. Cloud migration, generative AI, and automation aren't "future trends"; they're here now, and they're creating both opportunities and challenges. This section explores how these technologies fundamentally change issue and exception management. Like shifting tides in an ocean voyage, they can propel you forward or pull you under, depending on how you navigate.

Artificial Intelligence (AI):

Artificial Intelligence has quickly become one of the most valuable tools in modern GRC, not because it replaces people, but because it gives teams foresight they never had before. When AI and machine learning models analyze historical exception data, patterns emerge: which departments struggle most with compliance, which systems generate repeat issues, and even which seasons or business cycles drive elevated risk.

Practical Example: An AI system reviewing network logs and user behavior can spot unusual access patterns before they escalate into data-access exceptions. For a retail manager, that translates into catching inventory-system anomalies early, *before* they become PCI DSS violations, turning what could have been a regulatory fine into a routine operational fix.

AI also improves the front end of exception management. Instead of a flood of incomplete or poorly scoped requests landing on analysts' desks, automation can triage intake: gathering required details, assigning risk scores, and routing submissions to the right experts.

Practical Example: An AI chatbot guides employees through the intake process, capturing complete context and documentation before a human GRC analyst ever has to intervene. Analysts spend less time chasing missing information and more time evaluating real risk.

Beyond intake, anomaly detection enables AI to sift through massive datasets, highlighting subtle control breakdowns or emerging risks that warrant attention. Content analysis tools can also review contract language or policy drafts, flagging compliance gaps before new systems or vendor relationships go live.

In all cases, the goal remains the same: **use AI to anticipate and prevent exceptions, not merely to clean up after they occur.**

Of course, AI itself introduces new risks. Generative models bring challenges around bias, data exposure, hallucinated outputs, and opaque "black box" decision-making. As discussed earlier in this book, these technologies must operate under clear ethical controls and human oversight. AI should augment judgment, not replace it, especially when high-risk decisions are involved.

In fact, AI has become its own growing exception category, reflecting the reality that powerful tools can also become compliance liabilities when governance fails to keep pace.

Where AI identifies potential risk, automation accelerates response/ This shortens the exception lifecycle and transforms GRC from reactive enforcement to proactive risk management.

Automation and Orchestration (Workflow and Control Enforcement)

By automating workflows within the exception lifecycle, teams can reduce manual reviews based on predefined criteria. These processes can compile audit evidence, notify approvers when a case needs attention, or even close a case when criteria are met.

Practical Example: An automation could be set up so that when a team submits proof of remediation, the exception case is automatically closed, significantly reducing the need for manual human intervention and speeding up the overall process.

RPA (Robotic Process Automation, software that mimics human keystrokes to handle repetitive tasks) already plays a role for many companies. Customer Service and Data Entry are just two areas where these tasks are becoming increasingly prevalent, and GRC can leverage both. Other aspects of GRC where automation is playing a bigger role include evidence gathering and reporting.

One of the most exciting developments over the last few years is Automated Control Enforcement. Automated control enforcement (such as policy-as-code or configuration baselines) reduces exceptions but still requires human validation. This allows rules to be implemented directly in systems to prevent unauthorized actions, such as blocking unapproved software installs or automating access revocation. However, relying too heavily on automation for high-risk approvals or case closures can create critical blind spots and lead to significant audit challenges.

Automation helps teams work faster but introduces risk. Without regular oversight, over-reliance on automated enforcement can breed complacency, with teams assuming every control is self-healing and ceasing to monitor. Poorly scoped automation can incorrectly close or approve cases at scale. Automated workflows should still enforce segregation of duties and least-privilege principles to prevent system-wide approval errors.

At a global SaaS firm, a skeptical manager became an advocate for automation after seeing her team's manual workload drop by half. Her willingness to learn and to have the conversation helped spark a culture of continuous experimentation across the department.

Automation thrives in cloud environments, where scale and integration amplify both benefits and risks.

Cloud Environments:

Cloud environments aren't new, but how we use them is evolving. Cloud-native tools enable real-time dashboards and cross-region visibility into exceptions, giving managers a single, accurate picture of risk. Using Infrastructure-as-Code allows teams to embed compliance checks into deployment pipelines, preventing exceptions at the source.

Practical Example: When Infrastructure-as-Code (where infrastructure configurations are written and managed as software scripts) is deployed, teams gain built-in traceability and accountability for system changes. Every modification becomes documented evidence showing exactly what changed, who approved it, when it occurred, and why. This automatic process creates a reliable audit trail without manual reconstruction.

Cloud computing offers unmatched scale and agility for deploying GRC. It also lets you centralize security and compliance data. Many cloud-native services and configurations can be built to automatically enforce compliance, following the principle of compliance by design. This principle, often called security by design or compliance by design, integrates controls and audit evidence into the system configuration itself.

As organizations expand their cloud use, often across multiple vendors, complexity and risk increase. Managing access and configurations in more than one environment makes it easier for settings to drift or for IAM (Identity and Access Management) policies to become inconsistent between platforms. Auditors focus on these issues under SOC 2's Monitoring Activities (CC4), which emphasize continuous evaluation of control effectiveness. Routine scanning paired with centralized logging helps teams detect configuration drift early, before it escalates into formal audit findings.

These technologies are not isolated; they often work together. AI is sometimes used to analyze cloud data, and automation can orchestrate cloud security controls, all of it managed via a SaaS GRC platform.

APIs: The Middleman of Connections:

Application Programming Interfaces (APIs) serve as the connective tissue of modern digital systems, acting as standardized "digital messengers" that enable software platforms to communicate and exchange data. In the context of GRC, APIs are fundamental enablers for leveraging modern technologies like AI and automation. They move data from operational systems (e.g., identity management, cloud environments, threat feeds) into GRC platforms, enabling automated monitoring, real-time risk assessment, and streamlined exception workflows.

Integrations built on these APIs enable GRC tools to orchestrate actions, revoking access after AI-flagged policy violations or populating dashboards directly from source systems, reducing manual effort while improving speed and accuracy.

Why APIs Matter:

While offering immense benefits in efficiency and effectiveness, APIs and integrations also introduce their own set of GRC-related risks. Key risks include broken object-level authorization (BOLA), improper authentication, and excessive data exposure. Each of these risks opens new attack vectors. Furthermore, data integrity risks arise if integrations are not robustly designed, leading to inaccurate information flowing into GRC systems and potentially flawed risk assessments.

Effective GRC requires strong API security governance (policies that control how systems share data), clear data contracts, and continuous monitoring to keep digital connections secure, reliable, and compliant. APIs streamline processes, but poorly governed APIs can quickly become the weakest link in your security chain.

These connections create incredible power, but without strong governance, they also create new exception categories. Let's look at how to stay ahead of that curve.

Practitioner's Note: Embrace but verify

Leverage AI and automation to scale exception management, but only with human oversight. The fastest path to reputational damage is trusting a "black box" without understanding its limits.

What Not to Do:

Lose Focus.

- Don't treat AI or the cloud as silver bullets. They amplify both good and bad processes.

- Don't skip governance. Automate poorly, and you'll multiply errors, not fix them.

- Don't ignore new regulatory landscapes (e.g., the AI Act and SEC cyber rules). Today's innovation is tomorrow's compliance challenge.

As a manager, you don't need to be an AI expert, but you must understand how these technologies can fundamentally change your GRC team's efficiency, effectiveness, and pace. Work with your team to identify where automation can reduce manual churn and where AI can help predict problems before they become crises.

How to Futureproof Your GRC Practice

Futureproofing is often a losing game. Technology advances too quickly, and the returns on static defenses diminish fast. This section isn't about building something that never needs to change. It's about positioning yourself and your team to adjust more easily. Why? Exception management isn't static. To stay relevant, you must anticipate change and design for adaptability.

Build Agility into Your GRC Framework.

While 'Agile' is a specific project management methodology, the *principles of agility* are what GRC teams should adopt. Embrace sprints for GRC, test minimum viable controls, gather feedback, iterate.

Another strategy is to review corporate policies annually. I've seen places go several years between reviews, and as fast as technology shifts, this can be disruptive. Long gaps create policy debt. When reviews finally happen, teams face a scramble, mass updates, re-approvals, re-training, and audit panic, all at once. Regular updates prevent that pile-up and protect your people. This is especially effective when combined with continuous monitoring and auditing to identify deviations as they occur.

Continuous Metrics Maturity

Metrics evolve just like threats. Review KPIs quarterly to confirm they still measure what matters. A dashboard designed for yesterday's risks can quietly mislead leadership today, so continuous recalibration keeps improvement on course. What you measure defines what improves. Revisit KPIs regularly to ensure they reflect current risks and business value.

Investing in your team's skills delivers incredible returns. Train teams on cloud security, AI ethics, and automation risks so they can handle the changes driven by rapidly evolving technologies. Developers should receive training in privacy engineering awareness (aligned to NIST SP 800-160 Vol. 1 and 2). Cross-training with IT, legal, and DevOps builds holistic risk understanding and proactive capability. One final skill to develop is storytelling; GRC leaders must be able to translate risk into business impact quickly.

Strategic Risk Management and Scenario Planning

Strategic risk management is another way to build agility. Some consider it just a matter of aligning with evolving regulations. Still, if your team actively prepares for future events, you will be able to handle situations that arise. Conduct workshops to imagine scenarios that could occur, such as quantum computing breaking current encryption, and prepare GRC responses. For more details, refer back to Chapter 1's Tech Deep Dive on quantum computing. In particular, prepare for post-quantum risks, where future quantum computers could render today's RSA and ECC encryption obsolete.

Futureproofing becomes real when teams practice it, not just plan it. Theories become lessons when we see them tested in the real world. The following examples show how preparation pays off.

At a manufacturing firm, agile policy reviews caught GDPR changes early, avoiding a $100K penalty, and proving that adaptability pays.

Takeaway: The next generation of GRC leaders cannot just be mere rule enforcers. Leaders must also be risk translators and business enablers. Build systems and teams that thrive amid uncertainty.

Case Studies: Learning from Experience

Biotech Example: The Proactive Audit and How a Strategy Meeting Two Months Before Saved a Deal

The Situation: A startup that was a disruptive force in biotechnology was trying to increase market share for its SaaS product rapidly. A major university with enormous influence was willing to sign a seven-figure deal, but *only* if the company could attain a HITRUST certification.

Futureproofing in Action: Two months earlier, the IT team met with other departments to discuss pursuing SOC 2 attestation and ISO 27001 certifications. The team reviewed the work that needed to be done and the likely benefits of each. A manager from software development asked about other certifying bodies, and notes were taken on the processes each would follow, and how to prepare for multiple certifications at once. This proactive discussion positioned them perfectly when the HITRUST requirement emerged.

The Outcome: As soon as the possibility of the deal was raised, the meeting documentation was used to outline a rapid-response plan to move forward. Senior leadership was astounded that a well-considered plan showing the options the team had was delivered the next day, and a few months later, the deal was signed.

Childcare Example: Not Thinking Ahead Means Falling Behind

The Situation: A childcare provider experienced rapid growth and did not evolve beyond the ERP it had used for years. The software was so outdated that the vendor had only two remaining staff who knew how to support it. The ERP provider informed the childcare provider that it would have to switch to another product or platform within 6 months.

Futureproofing (in)Action: Despite the clear deadline, two months passed in meetings without a concrete plan. When they finally began evaluating new software solutions in earnest, they encountered a series of roadblocks. Their first choice lacked essential services, the second failed to meet security requirements, and the third proved prohibitively expensive. This reactive scramble left them with only six weeks before the legacy system's end-of-life, forcing a desperate choice: adopt the most expensive option or face shutting down operations.

The Outcome: The company spent far more than it would have if it had been proactive and remained agile in its processes.

Behind both stories is the same truth: agility determines resilience. To cultivate it, start small.

Practitioner's Note: The Human Factor and Adapting as a Team

Change isn't easy. Rapid shifts in tech and compliance can cause fatigue and resistance. The best GRC leaders recognize this, communicating transparently about why changes are happening and involving teams early. Celebrate quick wins and incremental progress; adaptability isn't just technical, it's cultural. Cultural adaptability begins with awareness. The simplest way to build that awareness is to test your own process.

Quick Win: Run an Exception Management Health Check

Before you can improve, you need to know where you stand. A simple, self-administered health check can quickly identify areas for improvement in your current IEM process. A health check is a lightweight way to benchmark your program and identify opportunities for immediate improvement, without waiting for the next audit. Doing this regularly can prevent major pain points down the road.

For organizations in scope for CMMC, tie each health-check finding back to the relevant NIST SP 800-171 control to keep evidence collection audit-ready.

This self-audit closes the loop on Chapters 3–6: tying lifecycle, metrics, and continuous improvement into one practical habit. Like a car's annual tune-up, this health check flags gaps early.

Use the checklist below to conduct your own internal health check.

Takeaway: Schedule your first self-audit this month, pick one IEM case. Walk it through the checklist, and share results with your team.

Practitioner's Note: Start Small, Show Results

A health check doesn't need executive sponsorship to begin. Run one in your own area, share results, and build momentum for broader adoption. Small, repeated wins are the heartbeat of continuous improvement introduced back in Chapter 6.

Where to Learn More: Resources, Communities, Continued Growth

Continuous improvement doesn't end with this book. GRC professionals must remain lifelong students, expanding knowledge and networks. Find groups on LinkedIn, attend conferences, and keep learning.

Recommended Resources

Frameworks and Tools

- Frameworks: NIST RMF, ISO 27001, COBIT, FAIR (a model for quantifying risk in financial terms).

- Tools: Power BI, ServiceNow GRC, LogicGate, Archer for

dashboards and metrics.

- Publications: ISACA Journal, CISO Tradecraft Podcast, CISA Alerts.

Communities and Conversations
- ISACA, ISC2, IIA, GRCI (Governance, Risk & Compliance Institute). These groups offer peer learning and local events.

- LinkedIn GRC Groups: Active discussions on emerging trends.

- Slack/Discord Security Communities: Informal peer support and rapid Q&A.

- Online communities like Reddit's r/GRC, and r/cybersecurity.

Continuous Learning Habits
- Subscribe to industry newsletters TechCrunch (tech trends), Krebs on Security, Sprinto's blog, CISO Talk, and other legal/compliance blogs.

- Podcasts: 'Risk Is Our Business' for 2025 trends, 'GRC & Me' for economic insights. Communities: GRC World Forums for global discussions.

- Block monthly time for skill-building (AI ethics, cloud certifications).

- Revisit metrics quarterly. Do they still align with evolving risks?

- **Certificates and Academic Resources**

- Certifications: CISSP, CISA, CRISC, CISM, CIPP, CGRC, CCEP.

- Attend conventions, conferences, and webinars. Industry events are invaluable for networking and learning.

- Academic Resources: University courses, research papers.

Continuous learning keeps your program resilient long after this book ends.

Conclusion

GRC exception management is no longer about static compliance; it's about adaptive resilience. The organizations that succeed are those that harness new technologies responsibly, are flexible, and commit to continuous improvement.

A team I was on conducted quarterly reviews, and our manager noticed that exception closure rates were lagging. Instead of just noting the metric, she launched a two-week sprint focused on closure, resulting in a 40% improvement and fewer audit findings next cycle.

Prepare for the future by harnessing new technologies responsibly, building flexible frameworks, aligning metrics to business value, and committing to continuous learning.

Now that we've learned what to do, how to do it, and how to prepare for the future, let's take a look at some of the experiences that others have had. Learning from our own successes and failures is valuable, but learning from the mistakes and wins of others makes the tuition much more affordable.

The best GRC programs aren't the biggest or most expensive; they're the most adaptable. Start today: review one process, try one improvement, and see how far you can go. With the groundwork laid for continuous improvement, our next chapter will dive deep into troubleshooting and FAQs, so you can avoid costly mistakes and accelerate your GRC success.

Manager's Perspective:

This chapter was focused on futureproofing GRC. While the term 'futureproofing' is imprecise, you can prepare yourself and your team for tomorrow's challenges by taking the right action today. You and your team will adapt or fall behind. With adaptability as your compass and continuous learning as your fuel, you're ready for the final stage. It's time to turn lessons into lived practice.

Small Team Tip: Continuous improvement doesn't require a big team. Schedule a quarterly "health check" to review what's working, what isn't, and one new tool or automation to try. Join online communities or user groups to learn hacks from other lean GRC teams. Peer support is a force multiplier.

End-of-Chapter Checklist

- Conduct a "health check" of your current exception management process.

- Identify at least one area to automate or improve with modern tools (AI, cloud, workflow software).

- Review industry resources, professional communities, or forums for new best practices.

- Set a quarterly goal for process improvement or innovation in your GRC practice.

- Document lessons learned and share them with your team or network.

Like evolving from a rigid map to a dynamic GPS, your GRC journey, fueled by this book's tools, positions you to navigate tomorrow's risks with confidence. Adaptability, continuous learning, and a culture of resilience are what keep your program relevant as technology, regulations, and business priorities shift.

But even the best-prepared teams hit roadblocks. Questions come up, processes stall, and exceptions don't always fit neatly into a framework. That's where we turn next. Chapter 8 tackles the most common challenges, FAQs, and troubleshooting scenarios so that you can move from theory to practice with fewer detours.

Chapter 8: FAQ/Troubleshooting: Navigating GRC Exceptions Like a Pro

If You Only Remember One Thing: Start with This FAQ, Then Build Your Own.

This chapter is your emergency toolkit for navigating common GRC challenges, from resistant stakeholders to overflowing backlogs. These FAQs and troubleshooting examples draw from real-world GRC headaches I've tackled. I'm sharing them with you here – all of the quick, actionable fixes and none of the stress. Each answer keeps it under 300 words, focusing on practical steps with metaphors and anecdotes to make it stick for non-experts in SaaS, retail, or healthcare. Let's dive in.

Strategic & Management Buy-In

How do I get management buy-in for a robust exception management program?

Securing buy-in is about clear communication of value. Tie your program's success to the big picture. Start by linking IEM to business wins. Reduced risks mean fewer breaches (and fines), while enabling safe innovation keeps revenue flowing. At a SaaS firm I consulted, we showed leadership how exceptions sped up a cloud migration without audit failures, framing them as "detours that get us to market faster." Use data where possible; you can even highlight potential fines versus savings from proactive fixes.

My leadership sees exceptions as failures. How do I change that perception?

You can start by reframing exceptions as necessary process adjustments in a dynamic environment, not fundamental failures. Exceptions are signs of a dynamic world, urgent needs, and tech limitations. Exceptions offer teams a chance to improve processes.

At a retail chain I advised, leadership viewed an exception for data encryption as a planning failure until we shared a recent case. A documented exception during the holiday rush allowed the company to meet business goals, while compensating controls helped to prevent greater risk. This showed that IEM is a place where we can take controlled risks to avoid system failure. Perspective is key. Emphasize that exceptions are opportunities: analyze patterns to fix systemic gaps, such as outdated policies.

How do I demonstrate the ROI of effective exception management?

ROI shines when you quantify the "invisible saves," like averting a data leak. Track metrics like reduced audit findings (e.g., from 15 to 5 per cycle), faster remediation (cutting weeks to days), fewer recurring exceptions (down 40%), and avoided losses (e.g., fines via monitored exceptions). In a healthcare EMR project, we showed leadership how IEM slashed HIPAA risks, saving $500K in potential penalties while speeding telehealth launches. You can also use dashboards to compare pre-/post-IEM costs, like breach cleanup at an average of $4M vs. proactive fixes at a fraction of that.

What are the three most critical metrics I should track to prove the health of my exception management program?

Focus on **Closure Rate** (exceptions closed/exceptions requested), **Average Days Open** (average length of time cases have been open), and **Exception Recurrence Rate** (how often the same issue comes up). These metrics go beyond dollars; they deliver operational resilience and audit readiness.

Operational Challenges & Process Management

What should I do if my exception backlog is growing out of control?

An out-of-control backlog isn't a personal failure; it's a growth signal. As exception programs mature, requests tend to surge faster than resources grow. The solution is to regain control in layers rather than brute-forcing the pile.

Start with risk triage. Use a simple impact-likelihood matrix and focus first on high-risk items, things like active vulnerabilities, regulatory exposure, or customer data risks. Low-risk requests can often be automated or batch-approved through workflow rules.

Next, create a rhythm rather than chaos. A weekly 30-minute backlog sprint, dedicated time to review, close, or escalate cases, prevents exceptions from stagnating. Finally, hunt the root causes. Recurring vendor problems, outdated policies, or unclear intake guidance create repeat exceptions. Training updates or minor policy tweaks often reduce request volume more than adding staff.

I saw this play out during a manufacturing audit where a neglected safety-exception backlog led directly to fines. We stabilized the program by categorizing the cases—closing about 20% that were outdated and escalating 30% that represented genuine high risk—then using pattern analysis to prevent recurrence. Within one quarter, the backlog was under control, and audit pressure dropped.

How do I handle "shadow exceptions" (exceptions not formally reported)?

Shadow exceptions are unaddressed risks. Bring them to light by making formal reporting processes clear and inviting (and not punishing teams for reporting risk). Improve the intake process with user-friendly forms (e.g., mobile-friendly tickets) and explain "why" it matters. Foster transparency by sharing success stories where reported exceptions averted fines. In a tech startup, informal workarounds led to audit failures. We fixed it with "no-blame" campaigns and quick-response office hours.

What do we do if a vendor, legacy system, or third-party won't/can't comply?

When a vendor or legacy system can't meet requirements, your job is to manage—not eliminate—the risk. Start by documenting your due diligence. Keep records of conversations, remediation requests, and any obstacles encountered. This protects your organization and demonstrates responsible oversight. Use contracts as leverage where possible. Set expectations for upgrades, remediation timelines, or exit clauses.

For systems that genuinely cannot be replaced quickly, approve time-bound exceptions supported by compensating controls, such as network segmentation or enhanced monitoring. I saw firsthand that vendor delays resulted in fines because escalation came too late. We corrected course by formalizing escalation timelines and adding dashboard-based tracking, which allowed leadership to spot vendor risk before it became enforcement action.

How do I ensure consistency in evaluation and approval when multiple teams are involved?

Inconsistency undermines credibility, mainly when different teams apply different standards. Without shared criteria, approvals can feel arbitrary, creating friction between the business and security teams. The fix is alignment around structure, not personalities. Use a scoring matrix to rate risks on impact and likelihood with predefined thresholds, so everyone is speaking the same risk language.

Define a clear approval hierarchy in which the business validates the operational need and security evaluates the risk. Reinforce this through periodic workshops or tabletop exercises to calibrate decision-making across teams. Finally, centralize visibility using a shared tracking tool like Archer or a similar workflow system so exceptions don't live in silos. I saw this firsthand at a multinational SaaS organization where scattered approvals were causing constant disputes. Once we implemented a shared risk matrix and standardized workflow, approval conflicts dropped by 50%.

Our teams are resistant to using the new exception process/tool. How can I increase adoption?

Resistance to new processes is common; it's human nature. Teams adopt tools when they see personal value, not when they're instructed to comply.

Start by choosing intuitive tools and actively share early wins, such as showing how the new workflow cut remediation time by 40%. Keep training short and practical, using real use cases rather than slide decks. Secure visible leadership support. Have executives explaining the "why" at all-hands meetings so the process feels mission-driven rather than imposed. Most importantly, listen. Address pain points, collect feedback, and iterate quickly so teams see that their input shapes the process. In one healthcare rollout, resistance faded once quick wins, such as automated reminders that prevent fines, became visible to everyday users.

The business team needs a quick exception for a deadline, but the security team sees high risk. How do we resolve this common conflict quickly?

Emphasize the role of robust Compensating Controls as a compromise· The decision is a Risk Acceptance process, not a security vs. business veto. Use the defined Approval Hierarchy (e.g., CIO/C-suite) as the tiebreaker to show it's an executive decision, not a GRC-team roadblock. Convene the parties in a 15-minute huddle. You set the tone: 'Let's find a safe path forward.'

Scope & Definition Clarity

What's the difference between an exception and an issue, and why does it matter?

This is one of the most common points of confusion in GRC, and getting it wrong has real consequences. An exception is a deliberate, documented, time-limited deviation from a control that leadership has consciously approved. An issue is an unexpected control failure, incident, or exposure that requires immediate response actions.

The distinction matters because the workflows are entirely different. Exceptions trigger remediation planning. Issues trigger incident response. When teams blur that line, problems stall instead of being fixed. I saw this firsthand at a retail company where a live data leak was treated as an "exception" rather than an issue. Response was delayed while paperwork was processed, and that delay directly contributed to regulatory fines. If you need a fast field test: Deliberate and documented? Exception. Unexpected and harmful? Issue. (For a deeper explanation of this distinction, see Chapter 2.)

When is it appropriate to deny an exception request?

Deny when risks clearly outweigh the business benefits; it's about protecting the organization's integrity. Criteria for denial: Unacceptable risk (e.g., no compensating controls), critical non-compliance (HIPAA/GDPR breaches), or no remediation plan. In a tech firm's rush, we denied a vendor's access request due to insufficient monitoring, averting an exploit of a vulnerability. Communicate effectively: "Here's why (risk details), alternatives (e.g., compliant tool), and appeal path."

How do I enforce a culture in which exception owners actively work toward remediation, rather than treating the exception as a 'forever waiver'?

Building an authentic remediation culture is hard, especially without strong executive sponsorship. You may not be able to force behavior across the organization, but you can shape accountability within the exception process itself. The most effective lever is making time-bound remediation non-negotiable in every approval.

Frame exceptions as a loan that must be repaid, not as permission slips to ignore controls. Every approval should include an expiration date, a clear remediation owner, and a documented next step toward closure. Even when broad culture change is slow, this structure prevents "forever waivers" and keeps exceptions moving toward resolution rather than quiet abandonment.

Resourcing & Scaling

My GRC IEM team is small. How can we effectively manage exceptions with limited resources?

Small teams succeed by becoming ruthlessly efficient. Prioritization and automation matter more when every headcount decision has real tradeoffs. Focus first on automating the lifecycle wherever you can—simple triage rules, reminder workflows, or automatic closure of low-risk cases once remediation is verified. Tool sophistication matters less than consistency.

Many teams start effectively with accessible options like Google Sheets or open-source platforms such as CISO Assistant, OpenGRC, eramba Community Edition, or SimpleRisk. Clear role assignments are just as crucial as tooling. Even a lean team benefits from knowing exactly who owns intake, review, and follow-up.

Pair this with lightweight dashboards (the example in Chapter 6 is a solid starting point) to keep priorities visible and prevent work from drifting into blind spots. I've seen this approach work firsthand. At a startup I supported, a three-person GRC team reduced its backlog by 60% simply by automating approvals and using dashboards to keep daily focus on risk-based priorities.

How can our exception management process scale as the organization grows?

Scaling your exception management program starts with building flexible foundations that don't collapse under volume. While adaptability will always remain your greatest asset, growing organizations benefit from embedding it within standardized structures such as clear scoring matrices, repeatable workflows, and well-defined escalation paths. As volume increases, automation becomes essential. Scalable GRC tools such as ServiceNow or LogicGate help centralize intake, route approvals, and track remediation without adding administrative burden.

Cross-training team members builds resilience so knowledge doesn't bottleneck as workloads expand. Continued growth depends on staying curious: review your metrics regularly and introduce new automation or workflow tools when your current processes begin to strain under scale. I saw this approach succeed at a SaaS firm that expanded from managing 50 to over 500 exceptions per month by centralizing intake and deploying AI-assisted reviews, allowing the team to scale volume without scaling headcount at the same rate.

How do I train new staff or teams on exception management?

New staff learn best when you start simple and build gradually. Begin with checklists that cover the fundamentals, including key definitions, submission workflows, and escalation steps, so no one feels lost on day one. From there, reinforce understanding through short microlearning sessions, such as 15-minute videos or tool walk-throughs. But the real learning happens through mentorship. Pairing new team members with experienced colleagues lets them see how exceptions are handled in real situations, using live examples rather than theory. After-action reviews and brief case studies tie everything together, helping lessons stick and reinforcing decision judgment. In my first InfoSec role, simple, quick-reference guides were essential to building confidence quickly, a principle that still holds for onboarding today.

Audit & Compliance Interactions

What do auditors look for when reviewing exceptions?

Auditors aren't trying to catch mistakes; they're validating that your process is real and followed. When reviewing exceptions, they look for gaps in execution, missing documentation, and instances where teams don't follow their stated procedures. The safest approach is to maintain clear, auditable records that document consistent evaluation, approval, tracking, and remediation for each exception. I saw this during a SOC 2 audit where strong documentation turned multiple potential findings into simple observations.

How can I prepare for an audit when I know we have many open exceptions?

A large number of open exceptions doesn't mean you're unprepared for an audit. It means you need disciplined visibility. Auditors don't expect zero open items; instead, they hope to see that risks are acknowledged, documented, and actively managed. Focus on keeping every case current. Update statuses, document ongoing remediation plans, and escalate high-risk items before the audit begins. Even when exceptions can't be fully closed yet, showing that ownership and progress exist goes a long way with auditors. I saw this during a finance audit, where proactively escalating open risks and documenting remediation plans turned potential significant findings into routine discussions.

What if a regulator or auditor disagrees with our exception process or a specific exception?

Disagreement with an auditor or regulator is more common than most teams realize. **It doesn't mean you've failed.** The goal isn't to win an argument; it's to demonstrate that your decisions were reasonable, documented, and by risk. Lead with facts rather than defensiveness. Walk through your risk assessment, compensating controls, and the decision rationale behind the exception. Remember that auditors have no visibility beyond what exists in your documentation, so transparency and clear explanation are your greatest tools. Invite dialogue instead of escalation. Many disputes are resolved once both sides understand the business context and the controls in place. During a GDPR review, we diffused initial objections using monitoring metrics and collaborative discussion, ultimately adjusting the process without receiving formal findings.

Quick Wins & Pitfalls: Fast-Track Your Exception Management Success

Quick Wins

- **Automate Intake & Approvals Where You Can**Set up a simple intake form and automate notifications or reminders. This speeds up triage, reduces lost requests, and makes the process "real" overnight.

- **Standardize Your Exception Criteria**Use a risk-based scoring matrix or a checklist for evaluating all exception requests. Consistency builds trust with auditors and stakeholders.

- **Centralize Documentation (granted and denied)**Store all exceptions, granted or denied, in one place (a spreadsheet, GRC tool, or even a shared folder for small teams). This makes audits smoother and keeps nothing hidden.

- **Communicate the "Why"**Explain to business users how the exception process protects them and the organization. Framing exceptions as risk-managed business decisions (not just "red tape") increases adoption.

- **Schedule Regular Reviews**Block 30 minutes every quarter to review open exceptions. Even a small, recurring review cycle prevents exceptions from becoming permanent.

Common Pitfalls to Avoid

- **Letting Exceptions Go Unreviewed**"Temporary" exceptions become permanent unless you actively schedule reviews and require closure or renewal.

- **Skipping Documentation**If it's not written down, it didn't happen. Lack of documentation is the #1 audit finding and the most common reason for compliance headaches.

- **Treating Every Exception the Same (low risk ≠ high risk)**Not all exceptions are equally risky. Don't waste time with approvals for low-risk requests, but never shortcut high-risk ones.

- **Ignoring Business Realities**A process that ignores operational constraints or business needs will be bypassed. Engage stakeholders early, and adapt the process as needed.

- **Confusing Exceptions with Incidents or Issues**Be clear about definitions. Mixing up exceptions, incidents, and control failures leads to mismanagement and audit confusion.

Pro Tip: If you only do one thing, automate reminders for exception review dates. Even the best processes fail if everyone forgets to follow up.

Build your own FAQ in a shared wiki. Every new pain point adds one new answer. In six months, you'll have a living knowledge base that is your culture.

Chapter 9: Deep Dives: GRC Battles Won and Lost

If You Only Remember One Thing: Let Others Pay Your Tuition.

G RC isn't learned from slides—it's learned when something breaks, and everyone looks to you for answers. The stories ahead aren't hypotheticals. They're lived moments where good intentions collided with real consequences. Each one captures what works, what fails, and what you can do differently before you end up learning the lesson the hard way. Consider this your tuition paid by experience, ours and others'.

The AI Exemption: Navigating Uncharted Policy Waters

Scenario: At a rapidly scaling SaaS company, what looked like a routine software request triggered a full-blown GRC conflict. Our real estate team needed to use Autodesk design software for site planning and layouts. Buried in Autodesk's updated licensing terms was a new clause stating that any uploaded drawings could be used to train their generative AI. That directly violated our internal policy prohibiting the sharing of company intellectual property with external AI models.

A formal exception request was submitted. Because of the IP implications, our severity assessment rated it **P0**, the highest risk category. That designation immediately escalated approval to the executive tier. Both a C-suite security leader and a Senior Vice President from the business unit were required to sign off.

Goal: The immediate objective was straightforward: resolve the request without compromising intellectual property while keeping construction plans moving forward. Beneath that, however, was a deeper issue. We lacked a shared understanding of who owned approval authority for high-severity exceptions. The process technically existed, but it was not operationally clear.

Action: Once the P0 rating was assessed, tensions rose quickly. Legal felt their sign-off should be sufficient. Security pushed back, insisting executive oversight was required, given the exposure risk. The real estate leader, focused on meeting deadlines, found themselves trapped between two immovable positions.

Voices sharpened. Professional frustration tipped into personal remarks. At one point, the business stakeholder quietly dropped from the call, later explained as a sudden "internet issue," though the timing left little doubt about the cause.

Sensing that the conversation was sliding past problem-solving and into territory that could leave real scars, I stepped in. We needed to reset before damage became permanent. I paused the debate and tried to pull everyone back to a single question: what outcome protected the company while still allowing the business to move forward?

After the meeting, it became clear that the real vulnerability was not Autodesk but internal ambiguity. I drafted a concise, step-by-step escalation workflow spelling out approval authority for P0 exceptions.

Every stakeholder received the document. I also followed up individually to rebuild alignment and trust.

Resolution: The exemption was approved in alignment with IP policy while still allowing the project to proceed safely. More importantly, the escalation guide became the standard reference for high-risk exception approvals across the company, preventing similar conflicts from reaching the executive level.

What began as a simple software request resulted in a permanent governance improvement.

Key Takeaway:

Even experienced leaders struggle when process ownership is unclear. Proactive documentation is not bureaucracy. It is the difference between collaboration and collision.

The ERP Modernization: A Costly Lesson in Neglected GRC

If the first story showed what happens when process clarity collapses, this next one reveals what happens when leadership buy-in never materializes.

Scenario: Early in my tenure as Head of Information Security at a distinguished law school, we launched a multi-million-dollar ERP modernization project intended to be transformational. Leadership envisioned a seamless transition to a major industry vendor to modernize operations and improve data management across decades of student records.

Almost immediately, cracks appeared. Key stakeholders weren't fully invested. A long-tenured lead developer viewed the new platform as a threat to his influence and quietly resisted adoption. Most worrying of all, security considerations were consistently pushed aside in favor of functionality and schedule pressure.

Goal: The institution's overarching goal was a successful, on-budget, and secure migration to a modern ERP system, streamlining operations and improving data management for current and historical student records. For me, as the Information Security head, the immediate goal became identifying and mitigating the significant data security risks arising from the vendor's practices, advocating for stronger protective measures, and, ultimately, protecting the institution's sensitive information and reputation, even as the broader project governance faltered.

Action: Our concerns became real when the ERP vendor suffered a significant breach. A compromised server was confirmed to have stored our institutional data, including historical student records with names, Social Security numbers, and other sensitive identifiers.

I worked with our Head of Human Resources to draft a formal letter demanding answers and documented assurances.

A month later, our fears were confirmed. The vendor sent spreadsheets containing portions of the same sensitive dataset by email, accidentally including an admissions staff member on the CC line. That person came to me directly, shaken and unsure of what to do next.

Between that incident, departments experimenting with the new platform without security coordination, and the ongoing internal resistance, trust across the project began to collapse.

Resolution: The convergence of vendor missteps, fractured communication, and internal resistance ultimately led to the project's end. Substantial amounts of money were lost to cancellation fees, sunk implementation costs, and months of staff stipends paid while systems were dual-running during the halted migration.

In the end, after hundreds of thousands in unrecoverable expenses, the law school stood exactly where it began, still in need of a new ERP, but newly aware of how costly weak governance can be.

Key Takeaway:

Leadership buy-in isn't optional. If a vendor demonstrates that they can't protect your data, **believe them and move on to someone who will**.

The "Shadow IT" Data Breach: Turning Crisis into Systemic Improvement

Scenario: At a large enterprise, the marketing department, under pressure to move faster, began using an unapproved cloud-based CRM tool. IT and security were never informed. Sensitive customer data was stored in the platform without review, approval, or security validation. From a GRC perspective, we had no visibility into where the data lived, how it was protected, or whether the tool complied with internal policies and regulatory requirements. We did not yet know it, but risk was already accumulating quietly.

Goal: Once the shadow IT surfaced, our immediate objectives were clear: identify the scope of the unauthorized platform, secure the data, and bring the activity back into compliance as quickly as possible. Beyond containment, we also needed to address the root cause. The long-term goal was to turn the incident into an educational moment for the business and to build mechanisms to prevent similar bypasses.

Action: The turning point came when a GRC analyst reviewing routine network logs noticed abnormal outbound data flows that did not align with approved systems. The investigation led directly to the unauthorized CRM tool. Data had been resident there for some time. Fortunately, there was no evidence of active exploitation, which brought a moment of cautious relief.

Containment had to happen immediately. We issued an emergency exception to acknowledge the tool's temporary status and bring it under governance control. That step allowed us to quarantine the exposed data, secure system connections, and initiate a full root-cause investigation.

At the same time, I had a difficult conversation with the head of marketing and executive leadership. The goal was not to blame. It was alignment. We had to make the seriousness of the violation clear while preserving relationships essential to future cooperation.

Resolution: Through steady coordination, we migrated the marketing team's operations onto an approved, secure, and compliant CRM platform, eliminating the exposure risk.

More importantly, we used the incident to drive permanent improvements. A cross-functional lessons-learned session brought transparency to what went wrong and why bypassing formal review seemed easier than following the process. That session directly led to policy revisions that clarified cloud tool usage, introduced a streamlined cloud-solution review pathway, and expanded GRC office hours to provide early consultation for business technology initiatives.

Key Takeaway:

Not all wins begin as successes. This incident highlighted the dangers of shadow IT while also showing how effective GRC can transform a crisis into lasting improvement. Done well, governance does not slow innovation. It makes safe innovation possible.

While shadow systems hide in plain sight, some risks persist because they've been visible for so long, they fade into the background, as the following story shows.

The Legacy System Audit Failure: Confronting Technical Debt

Scenario: At a long-established manufacturing company, factory floor operations depended on a mission-critical control system that had been in place for decades. It ran on an operating system that no longer received security patches and lacked meaningful audit logging. From a compliance standpoint, it failed modern industry requirements, including ISO 27001 controls and critical infrastructure security standards.

Everyone knew the system was old. What we had not fully admitted was that it had quietly become a permanent exception. Its reliability and the cost of replacement kept pushing remediation into the future, until the risk faded into the background.

Goal: Once the issue surfaced during a regulatory audit, the immediate objective was to respond to the crisis. We had to manage the financial penalties, deal with furious executives who felt blindsided, and begin building a viable plan to address the exposure.

Our longer-term goal was more demanding. We needed to dismantle the technical debt that had accumulated around the legacy system, apply compensating controls to reduce operational risk, and develop a modernization roadmap that leadership would commit to funding and executing.

Action: The turning point came during the audit close-out meeting when the system was cited as a major compliance failure. The reaction inside leadership was immediate. Frustration turned to anger as the implications became clear, both financially and reputationally.

The GRC team moved quickly into response mode. Our first step was uncomfortable but necessary: formally documenting the years-long exception and putting ownership where it belonged. That transparency brought visibility to a problem that had lingered unspoken for too long.

We worked with engineering and IT to develop a multi-year modernization roadmap and tie it directly into the company's strategic planning cycle so it could no longer be deferred. In the interim, we implemented compensating safeguards, including strict network segmentation around the legacy controls and enhanced monitoring of system activity to reduce immediate exposure.

Throughout the process, we maintained steady communication with leadership, reporting progress openly and honestly as funding, timelines, and remediation obstacles evolved.

Resolution: Over time, phased upgrades began replacing pieces of the legacy system, and the modernization roadmap moved from plan to measurable execution. What started as a regulatory failure reshaped leadership's view of technical debt as an operational risk rather than an abstract IT concern.

To reinforce this shift, the GRC team implemented a Legacy Risk Management framework requiring named ownership for inherited systems, documented remediation timelines, and mandatory compensating controls. No future exception could be labeled temporary without a defined resolution plan and reporting cadence.

The following annual audit reflected measurable improvement, validating both the controls put in place and the governance model supporting them.

Key Takeaway:

Costly failures force necessary conversations. Addressing technical debt requires visibility, ownership, and the willingness to confront long-ignored risks. When governance finally stepped into the light, what began as a compliance defeat became the foundation for lasting infrastructure resilience.

The GDPR/Privacy Policy Pivot: GRC as a Strategic Enabler

Scenario: A fast-growing e-commerce platform was preparing to expand into several European markets, a significant strategic move tied directly to revenue growth targets. During early planning, the reality of GDPR and regional privacy regulations set in.

Initial assessments from legal and business teams were discouraging. Compliance appeared to require significant system redesigns and lengthy delays, threatening to push the launch well beyond the company's planned timeline. Leadership began to see privacy requirements less as protection for customers and more as a regulatory wall standing between the business and market entry.

Goal: Our immediate goal was to help the company enter Europe on schedule without incurring regulatory penalties or legal risk. Long-term, we aimed to shift the organization's entire view of compliance. If the work was done correctly, privacy controls could become a market enabler rather than a barrier.

Action: It became clear that simply handing teams a list of GDPR requirements would stall progress rather than advance it. We needed shared problem-solving, not checklists. The GRC team organized a series of privacy-by-design workshops that brought legal, product, and engineering leaders into joint working sessions.

Instead of framing privacy as a set of restrictions, we reframed it as a design challenge. We worked through product workflows together, identifying areas where personal data could be reduced, anonymized, or removed entirely. For features that required short-term deviations, we used structured, time-limited exceptions with clear remediation expectations so development could continue without creating uncontrolled risk.

One of the most important shifts came when leadership began tracking compliance readiness as a business metric rather than a legal burden. Translating regulatory requirements into measurable delivery milestones gave executives visibility they could trust and teams a target they could work toward.

Resolution: The platform launched into its European markets on schedule and in full compliance with applicable privacy regulations. What began as perceived regulatory resistance became one of the company's competitive advantages. Privacy readiness helped accelerate approvals, shortened contractual negotiations with enterprise customers, and set the organization apart from competitors still scrambling to retrofit controls after launch.

Key Takeaway:

GRC can move organizations forward when it replaces rigid compliance checklists with collaborative problem-solving. Privacy by design did more than manage the risk. It enabled faster market entry, with greater confidence, and stronger long-term trust.

The "Audit Theater" Success: Transforming Dread into Readiness

Scenario: At a mid-sized professional services firm, the annual regulatory audit had become an event everyone dreaded. Every year followed the same exhausting script. Staff would scramble to assemble fragmented documentation at the last minute. Anxiety rose ahead of interviews. Leadership braced for difficult findings.

This reactive pattern had become routine. Confidence eroded. Minor audit deficiencies appeared year after year, not because controls were absent, but because preparation and documentation discipline were inconsistent.

Goal: Break the cycle entirely. We wanted to replace panic with preparedness and make readiness part of daily operations instead of annual triage. That meant strengthening documentation practices, building staff's confidence in their responses to auditors, and reducing recurring audit findings so leadership could rely on the firm's control environment rather than worry about it.

Action: Recognizing the pattern would not change on its own, we introduced an "audit theater" program built around quarterly mock audits. These sessions were structured to mirror real audit interviews and evidence requests, but the tone was collaborative rather than corrective.

During the first few sessions, nerves were obvious. Teams hesitated. Answers were tentative. Documentation gaps surfaced quickly. Instead of treating these moments as failures, we used them as learning points. We worked through documentation trails together, corrected gaps in real time, and coached staff on how to respond clearly and confidently to audit questions.

To support behavioral changes, we introduced a standardized audit-trail template so teams could maintain evidence continuously rather than rebuild it each year. We supplemented this with lightweight dashboards that enabled leadership to view readiness levels at any time, not just during audit season.

Resolution: When the next annual audit arrived, the shift was unmistakable. Documentation was complete and organized in advance. Interviews felt calm and professional rather than tense and reactive. Staff answered questions confidently because they had already practiced the process.

Audit findings dropped significantly, reflecting not only improved preparation but a stronger overall control environment. Equally important, the firm reclaimed time and energy previously drained by last-minute audit chaos. Leadership confidence improved substantially as readiness became visible and repeatable rather than speculative.

Key Takeaway:

Preparation changes culture. Regular internal audits and consistent evidence hygiene replace fear with confidence and turn compliance into a steady operational rhythm rather than a once-a-year emergency.

We later formalized these practices into quarterly control-health metrics to ensure that preparedness remained a habit rather than an initiative.

The Cleartext Password Crisis: Doing the Hard Thing

Scenario: While working as a security professional at one of the country's largest providers of family support services, I uncovered a failure so basic it was almost unbelievable. In one of our business departments, employees were storing operational system passwords in plaintext files. Those files sat in a plainly labeled desktop folder called "passwords."

There was no encryption, no access restriction, and no containment. Anyone who accessed one of those machines had immediate access to sensitive systems and, by extension, client data. When I raised the issue with my direct supervisor, the head of IT, the response was dismissive. The urgency I felt did not match the urgency I received. It became clear that the issue would not be addressed through normal channels.

Goal: My first goal was non-negotiable and straightforward: eliminate the security risk immediately. Leaving cleartext passwords on employee desktops was a near-guarantee of an eventual breach.

The harder objective was knowing how to force that action when my own leadership chain was unwilling to move. Escalating meant risking professional relationships and possibly my job, but failing to act threatened far worse consequences for the organization and the people we served.

Action: After several stalled conversations, I decided escalation was the only responsible option. I documented every finding in detail, including evidence of the exposed credentials, affected systems, policy violations, and potential breach impact scenarios.

I requested time directly with senior executive leadership and laid out the risk plainly, without cushioning the implications. I bypassed my immediate supervisor, fully aware of what that could mean for my working relationships and career stability.

Leadership did not hesitate once the gravity of the situation became clear. They authorized immediate remediation. The "passwords" folders were eliminated across departments, and secure credential management practices were implemented immediately.

Resolution: Less than a week later, corporate leadership initiated a previously scheduled, unannounced external audit. The scope included exactly the areas impacted by the password exposure.

Had the folders still existed, the resulting finding would have been catastrophic. Instead, the auditors found compliant controls where a disaster had nearly occurred.

What could have become a public and damaging failure became a narrow escape made possible only because action was taken when it was uncomfortable to do so.

Key Takeaway:

Fundamental security hygiene is never optional. Security professionals must be willing to escalate when leadership stalls, even at personal risk. Courage, paired with decisive action, prevents disasters long before technical controls ever do.

The Lingering Cloud Credentials: A Wake-Up Call for IAM

Scenario: Four years after leaving a fast-growing biotech startup, I received an unexpected call from a former colleague. They were struggling with an issue in a portion of their cloud environment that I had worked on extensively. Wanting to help, I agreed to take a look.

When I attempted to access the system, I assumed I'd be blocked immediately. Instead, my credentials worked.

The same administrative credentials I created during my tenure in 2020 were still active and still carried broad access rights. I sat there staring at the login screen, realizing I could view and modify systems I hadn't touched in years.

The chill was immediate. This was not just a technical oversight. It was a critical failure of identity and access controls.

Goal: My original goal was to help resolve the technical issue for a former teammate. That vanished almost instantly. The absolute priority became closing the security gap the moment it revealed itself. Stale credentials with elevated access represented a direct and ongoing risk to the company's infrastructure.

Beyond fixing the immediate problem, I wanted to ensure the team recognized what this meant. IAM gaps aren't limited to forgotten user accounts. Service accounts, embedded credentials, and workaround access paths can quietly persist long after employee offboarding processes conclude.

Action: Once the functional issue was stabilized, I explained what I had just uncovered and why it mattered. I carefully and clearly walked through the implications. Any attacker who discovered similar credentials would inherit the same level of system access I still had.

Together, we mapped where credentials could exist outside their standard identity system. I advised initiating immediate searches for orphaned service accounts and standalone cloud credentials not tied to employee identities. Every non-managed credential was flagged for removal, rotation, or full integration into centralized IAM oversight.

Where removal was not possible, we discussed isolating systems through network segmentation and layered authentication so that no single credential could provide unrestricted access across critical infrastructure.

Resolution: Although I was no longer responsible for their security posture, the discovery triggered action. The team expanded their credential audits and began strengthening offboarding controls beyond traditional user accounts.

What started as a friendly troubleshooting call became a blunt demonstration of how easily access gaps can persist when IAM hygiene does not extend to every pathway into the environment.

Key Takeaway:

Stale credentials are one of the most overlooked risks in cloud environments. Effective GRC reaches beyond employee offboarding to include service accounts, embedded credentials, and technical access paths. When these are neglected, vulnerabilities can remain wide open for years without detection.

The Accidental GRC Architect: From Side Project to Strategic Mandate

Scenario: At a rapidly growing tech startup, innovation moved fast while governance quietly lagged. Compliance was handled informally across teams as issues arose, with no centralized reporting or dedicated GRC function. Leadership had little visibility into overall risk exposure beyond reacting to problems as they surfaced.

Noticing these gaps as the company scaled, the CEO asked a technically skilled IT team member to take on a small side project. The assignment was modest: spend roughly one day a week creating basic compliance dashboards to give leadership clearer visibility. At the time, it was framed as an experiment to see whether even limited reporting could provide value.

Goal: The immediate goal was straightforward: bring order to fragmented compliance data and translate it into dashboards that leadership could understand. Beyond that, the effort carried an unspoken test. Could GRC prove, through visible results rather than theory, that it deserved dedicated investment?

Action: With minimal tooling and limited time, the IT team member began pulling data from IT operations, legal documentation processes, and business unit audits. A small set of practical metrics emerged: open findings, overdue policy updates, control gaps, and regulatory obligations without documented owners.

Simple dashboards began to take shape. Within the first few months, patterns became impossible to ignore. Recurring audit findings were clustered in predictable areas. Long-unowned policies surfaced. Certain regulations were only partially addressed across business functions.

What had previously been anecdotal concern suddenly became visual evidence. In leadership meetings, the dashboards replaced vague discussions with actionable clarity. The questions shifted from "Are we okay?" to "Why are these controls repeatedly failing, and what's our plan?"

As new requests for deeper reporting rolled in, it became clear that demand was growing faster than the original assignment could support.

Resolution: Within six months, the results spoke for themselves. Leadership elevated the part-time experimental role to a full-time GRC subject-matter expert position, with a mandate to formalize the program.

Control mapping was expanded across SOC 2 and the NIST Cybersecurity Framework, bringing consistency to audit preparation and remediation tracking. Reporting evolved into proactive policy reviews, structured risk assessments, and early GRC involvement in product development to embed security by design rather than retrofit controls after launch.

What began as a one-day-a-week experiment became the backbone of the company's compliance and risk governance function, driven not by mandates but by demonstrated value.

Key Takeaway:

Effective GRC programs are often built by showing value before demanding authority. Clear reporting and tangible insights transform compliance from an obligation into an asset, enabling individuals to grow into strategic leaders who strengthen organizational resilience.

Practitioner's Note: Know Your Frameworks & Train Your Team

Every GRC battle, win or lose, teaches resilience. Use these stories to spark dialogue, not blame, and build a culture where risks are faced head-on. From Chapter 1's GRC mindset to Chapter 8's FAQs, these battles show how resilience turns risks into results. Each case, whether a costly ERP flop or a GDPR triumph, offers lessons: document clearly, collaborate early, and act decisively. Start one improvement today. Run a health check, train a team, or revise a policy, and you'll see your GRC program soar. Across every battle, one truth remains: maturity comes not from avoiding mistakes, but from codifying what they teach.

Conclusion: Turning Insight into Impact

The Journey You've Taken

You have moved through the complete lifecycle of modern Governance, Risk, and Compliance, from the initial intake of an exception to the final dashboard report. Along the way, we have addressed the friction points that define modern business: the rush of cloud adoption, the persistent weight of "permanent" exceptions, and the emerging frontier of AI and automation.

If there is a single takeaway from these chapters, it is this: **Effective GRC is not a static rulebook.** It is a dynamic practice that balances technical rigor with the human elements of trust and accountability. It is a living discipline that adapts, includes, and empowers people at every level of the organization.

The New Mandate for GRC Leaders

Today's GRC leaders are more than policy enforcers. They're:

- **Risk Translators:** Converting complex technical vulnerabilities into clear business impacts that leadership can act upon.

- **Trust Partners:** Moving away from being "gatekeepers" to becoming collaborators who enable the business to move faster, safely.

- **Culture Architects:** Building an environment where reporting an error is seen as an opportunity for growth rather than a cause for punishment.

Key Takeaway: Your job is not to slow down innovation. It's to help your organization move forward, smarter and safer.

Continuous Improvement

The strongest GRC programs utilize:

- **Iterative Metrics:** Reviewing data as a weekly habit, not just an annual audit exercise.

- **Practical Policies:** Shaping rules around how work actually happens, rather than forcing work to fit a theoretical checkbox.

- **The "Bad News Early" Protocol:** Cultivating a culture where early transparency is rewarded, preventing small exceptions from becoming catastrophic failures.

Remember: Your maturity isn't measured by how few exceptions you have, but by how you learn from and resolve them.

Your Next Steps

1. **Run a Health Check:** Take a fresh look at your current exception management process. Where are the blind spots?

2. **Build One Dashboard:** Start small. Track what matters.

Share progress and keep it visible.

3. **Find Your Champions:** Allies in every business unit can model transparency and good process. Make them visible and celebrate their wins.

4. **Commit to Learning:** Stay curious. Regulations and technology change, your playbook should, too.

Final Encouragement

GRC is often invisible—until the day it saves the company, protects a customer, or helps a colleague sleep better at night. That is the essence of leadership. That is your impact.

Transparency builds trust. Trust builds accountability. Accountability drives continuous improvement.

Keep adapting. Keep improving. Build systems that protect not only your organization but the people and values behind it.

If you found this book helpful or have your own stories to share, please reach out to me on LinkedIn. Let's keep learning together.

The following section offers overviews of the major frameworks mentioned throughout this handbook.

Appendix: Frameworks Deep Dive

NIST SP 800-53

Scope & Audience

NIST SP 800-53 is the foundational security and privacy controls catalog for U.S. federal agencies and many government contractors, though it's widely adopted in the private sector as a best practice. The framework guides organizations in selecting and implementing appropriate security controls to protect information systems.

Where Exceptions/Issues Most Commonly Occur

Most common exception requests involve legacy systems unable to meet newer control requirements, resource limitations that prevent timely implementation, or technical constraints (e.g., outdated encryption, unsupported multi-factor authentication).Frequent pain points: System & Communications Protection (SC), Access Control (AC), and Audit & Accountability (AU) controls.

Key Controls or Requirements to Watch

- **AC-2 (Account Management):** Exceptions often needed for non-standard accounts or vendor integrations.

- **SC-13 (Cryptographic Protection):** Legacy applications

may not support modern encryption algorithms.

- **AU-2 (Audit Events):** Logging requirements may be impossible to meet on some older platforms.

Quick Tip:

If you only remember one thing about NIST 800-53: *Document your rationale and risk mitigation steps for every exception, auditors expect a clear "why" and a plan, not just a waiver.*

ISO 27001

Scope & Audience

ISO/IEC 27001 is the international standard for Information Security Management Systems (ISMS). It is used by organizations worldwide, across all sectors, to establish, implement, and maintain risk-based security programs.

Where Exceptions/Issues Most Commonly Occur

Exceptions typically arise when organizations cannot fully implement one or more "Annex A" controls due to business process constraints, legacy technology, or integration with external partners.Fr equent pain points: Access control, asset management, and supplier relationship controls.

Key Controls or Requirements to Watch

- **A.9 (Access Control):** Challenges with legacy apps or third-party integrations.

- **A.12 (Operations Security):** Older infrastructure may not support all monitoring or logging requirements.

- **A.15 (Supplier Relationships):** Third-party risk management controls can be hard to enforce fully.

Quick Tip

If you only remember one thing about ISO 27001:*Every control not implemented must have a documented justification and an accepted risk, reviewed at least annually.*

PCI DSS

Scope & Audience

PCI DSS is the Payment Card Industry Data Security Standard, applicable to any organization that stores, processes, or transmits credit card data. It is mandatory for merchants and service providers handling payment card information.

Where Exceptions/Issues Most Commonly Occur

Exceptions are most often requested for legacy systems or point-of-sale devices that can't meet strict encryption or authentication requirements, as well as for compensating controls in highly customized environments. Frequent pain points: Encryption, multi-factor authentication, vulnerability management.

Key Controls or Requirements to Watch

- **Requirement 3 (Protect Stored Cardholder Data):** Legacy storage solutions may not encrypt data properly.

- **Requirement 8 (Identify and Authenticate Access):** Difficulties with enforcing MFA on older systems.

- **Requirement 11 (Vulnerability Management):** Automated scanning/patching may not be feasible everywhere.

Quick Tip

If you only remember one thing about PCI DSS:*Compensating controls must be formally documented and shown to be as effective as the original requirement.*

HIPAA

Scope & Audience

The Health Insurance Portability and Accountability Act (HIPAA) Security Rule applies to healthcare providers, insurers, and any business associates handling protected health information (PHI) in the United States.

Where Exceptions/Issues Most Commonly Occur

Most exceptions are related to legacy clinical applications, medical devices, or integrations that can't support full security controls.Fre quent pain points: Encryption, access control, audit logging, device security.

Key Controls or Requirements to Watch

- **164.312(a) (Access Control):** Unique user identification and automatic logoff are often difficult in shared device settings.

- **164.312(e) (Transmission Security):** Older medical devices may not support encrypted transmissions.

- **164.308(a) (Security Management Process):** Risk analysis and risk management documentation are perennial gaps.

Quick Tip

If you only remember one thing about HIPAA: *"Addressable" standards must still be considered and justified, lack of action always requires a written rationale.*

CMMC (Cybersecurity Maturity Model Certification)
Scope & Audience

The CMMC framework applies to all defense contractors and subcontractors within the Defense Industrial Base (DIB) that handle sensitive, unclassified government information. Compliance is tiered and mandatory:

- **Level 1 (Foundational):** Applies to companies handling **Federal Contract Information (FCI)**. Requires an annual self-assessment.

- **Level 2 (Advanced):** Applies to companies handling **Controlled Unclassified Information (CUI)**. Requires a triennial C3PAO (Third-Party) assessment for most contracts and is based on the 110 controls from NIST SP 800-171.

Where Exceptions/Issues Most Commonly Occur

Exceptions and issues frequently stem from misinterpreting the CMMC's strict, non-negotiable approach to the handling of CUI.

Frequent pain points: Scoping mismanagement, PO&M disallowance, an accurate SSP with certifiable evidence, MFA, access control, and configuration management.

Quick Tip

If you only remember one thing about CMMC:

Shrink your scope! Isolate CUI onto a dedicated, minimal "enclave" (separate systems, network, and personnel) to drastically reduce the number of controls you must implement and the overall cost of certification.

SOX

Scope & Audience

The Sarbanes-Oxley Act (SOX) applies to U.S. public companies, focusing on internal controls over financial reporting (ICFR). IT controls relevant to financial systems are also in scope.

Where Exceptions/Issues Most Commonly Occur

Exceptions often arise when legacy financial systems can't meet segregation of duties, audit logging, or access review requirements.F requent pain points: Change management, privileged access management, audit trails.

Key Controls or Requirements to Watch

- **Access Controls:** Limiting access to financial systems/data to authorized users only.

- **Change Management:** All changes to systems impacting financial data must be logged and reviewed.

- **Audit Logging:** Complete, tamper-resistant logs must be maintained and reviewed regularly.

Quick Tip

If you only remember one thing about SOX:*If a control can't be implemented as designed, document compensating controls and be ready to prove effectiveness to external auditors.*

GDPR

Scope & Audience

The General Data Protection Regulation (GDPR) governs data privacy for EU residents and applies globally to organizations that process their personal data.

Where Exceptions/Issues Most Commonly Occur

Common exceptions relate to data subject rights (erasure, access), consent management, and cross-border data transfer controls, especially when legacy systems or third-party vendors are involved.Frequent pain points: Data retention, consent tracking, encryption, international transfers.

Key Controls or Requirements to Watch

Article 5 (Principles): Lawful, fair, transparent processing and data minimization.

- **Article 32 (Security):** "Appropriate" technical and organizational security measures.

- **Article 17 (Right to Erasure):** Some legacy systems make data deletion technically difficult.

Quick Tip

If you only remember one thing about GDPR:*You must demonstrate ("show your work") how exceptions are justified, risks managed, and data subject rights upheld, paperwork is not optional.*

Additional Major Frameworks:

ISACA/COBIT

What it is:

COBIT (Control Objectives for Information and Related Technologies) is a comprehensive framework for IT governance and management, developed by ISACA. It helps organizations align IT goals with business objectives and manage risk, compliance, and value delivery.

Who uses it:

Enterprises of all sizes, especially those needing to formalize IT governance, risk, and compliance processes.

One key insight:

COBIT is best used as a "bridge" between technical teams and business leaders, ensuring IT delivers measurable business value.

NIST Cybersecurity Framework (CSF)

What it is:

The NIST CSF is a high-level, voluntary framework for managing and reducing cybersecurity risk. It provides a flexible set of guidelines structured around five core functions: Identify, Protect, Detect, Respond, and Recover.

Who uses it:

Organizations of all sizes and industries, especially those in critical infrastructure or seeking a best-practice baseline.

One key insight:

The CSF is modular, use it as a "starter kit" for building or improving your cybersecurity program, then layer in more detail as you mature.

COSO (Committee of Sponsoring Organizations of the Treadway Commission)

What it is:

COSO's Internal Control-Integrated Framework is the gold standard for enterprise risk management (ERM) and internal controls over financial reporting.

Who uses it:

Public companies, financial institutions, and organizations needing robust risk management or compliance with SOX.

One key insight:

COSO helps organizations "connect the dots" between risk, control activities, and organizational objectives, think of it as a big-picture ERM map.

Cloud Security Alliance (CSA) Cloud Controls Matrix (CCM)

What it is:

The CSA CCM is a control framework specifically designed to secure cloud computing environments. It maps cloud security requirements to standards like ISO 27001 and PCI DSS.

Who uses it:

Organizations adopting cloud infrastructure or SaaS solutions, cloud service providers, and security assessors.

One key insight:

Use the CCM to quickly assess gaps in your cloud provider's controls and to justify exceptions related to shared responsibility in the cloud.

CIS (Center for Internet Security) Critical Security Controls
What it is:

The CIS Controls are a prioritized set of practical, action-oriented cybersecurity best practices. They're designed as "quick wins" to reduce risk across all organization sizes.

Who uses it:

Any organization looking to strengthen basic security posture, especially small to midsize businesses without a dedicated security team.

One key insight:

Start with the top 6 controls ("basic hygiene") before tackling advanced measures, this alone dramatically lowers common cyber risks.

ISO 31000 (Risk Management Guidelines)
What it is:

ISO 31000 is a global standard for risk management, not just for IT or cybersecurity, but for any organizational risk. It offers principles and guidelines for building an effective risk management framework.

Who uses it:

Any organization (private, public, or non-profit) looking for a systematic, enterprise-wide approach to risk management.

One key insight:

ISO 31000 emphasizes that risk management is everyone's responsibility, not just a function for compliance teams.

Glossary of Key Terms

This glossary clarifies core concepts from the book, focusing on exception and issue management. It's alphabetized for quick reference, with plain language definitions, metaphors, and examples to make GRC feel like a trusted co-pilot rather than a distant expert. Use it as a decoder ring for your own GRC journey, dip in during a coffee break, or share a term on LinkedIn to spark discussions.

Core Concepts

Access Control*Definition:* Policies and tools that ensure only authorized users can access specific data or systems.

Accountability Culture*Definition:* An environment where individuals and teams own risks, controls, and exceptions without fear of blame, like a team sport where everyone is responsible for defense.

Actionable Metric*Definition:* A measurement that sparks decisions or changes behavior (e.g., "average closure time," not just "exceptions filed").

Agility*Definition:* The ability of an organization to quickly adapt to changing risks, regulations, and business needs.

AICPA (American Institute of Certified Public Accountants)*Definition:* A professional organization that develops audit standards, including SOC reports.

API (Application Programming Interface)*Definition:* Digital "messengers" that let software systems talk to each other, automating control checks or data transfers.

Approval (Exception Request)*Definition:* The formal process of reviewing and accepting an exception request, often with multiple stakeholders.

Attestation*Definition:* A formal declaration that controls are in place and effective, standard in SOC audits.

Audit Trail Template*Definition:* A standardized document for recording who did what, when, and why, essential for passing audits.

Automated Control Enforcement*Definition:* Using technology to enforce rules or policies automatically (e.g., password requirements, access expirations).

Baseline*Definition:* The minimum standard or starting point for controls, policies, or system configurations.

Black Box Decisions*Definition:* Outputs from AI systems where the logic is hidden, making decisions hard to explain or audit.

Business Continuity Plan (BCP)*Definition:* A documented strategy to keep operations running during or after a disruption.

Certified Compliance and Ethics Professional (CCEP)*Definition:* Certification for professionals managing compliance and ethics programs.

Certified Information Systems Auditor (CISA)*Definition:* The ISACA certification for IT audit, control, and security.

Certified Information Security Manager (CISM)*Definition:* The ISACA certification for managing enterprise information security.

Certified Information Privacy Professional (CIPP)*Definition:* The IAPP certification for privacy laws and regulations like GDPR or HIPAA.

Certified in Governance, Risk and Compliance (CGRC)*Definition:* The (ISC)² certification for broad GRC knowledge and best practices.

Certified in Risk and Information Systems Control (CRISC)*Definition:* This ISACA certification is focused on IT risk management.

Change Management*Definition:* A structured approach to transitioning individuals, teams, or organizations to a desired future state.

Chief Information Security Officer (CISO)*Definition:* Senior executive responsible for information security and risk.

COBIT*Definition:* Framework for IT governance, aligning technology with business goals.

Compensating Control*Definition:* An alternative safeguard used when a primary control can't be implemented (e.g., extra monitoring if MFA isn't possible).

Compliance by Design*Definition:* Building compliance requirements into systems and processes from the start.

Continuous Improvement*Definition:* Regularly reviewing and refining GRC practices to match emerging risks and lessons learned.

Control*Definition:* Any process, policy, or technical measure used to manage risk (e.g., access reviews, encryption).

Control Owner*Definition:* The person or team responsible for maintaining a specific control.

Control Weakness*Definition:* A gap or failure in a control that exposes the organization to risk.

Cybersecurity Maturity Model Certification (CMMC)*Definition:* DoD framework for assessing cybersecurity practices in defense contractors.

Data Minimization*Definition:* Collecting and retaining only the data strictly necessary for business purposes.

De-Identification*Definition:* Removing or masking personal identifiers from data sets to protect privacy.

Denial (Exception Request)*Definition:* The act of rejecting an exception request due to unacceptable risk or lack of justification.

Due Diligence*Definition:* Thoroughly checking a vendor, system, or process for risks before approval.

Escalation Path*Definition:* A predefined process for raising high-risk or unresolved exceptions to higher authorities.

Exception*Definition:* A documented, approved, temporary deviation from policy, with risk mitigations and a plan to return to compliance.

Exception Creep*Definition:* When multiple "temporary" exceptions are never closed, leading to weakened controls and increased risk.

Exception Lifecycle*Definition:* The five steps every exception should follow: intake, evaluation, approval/denial, documentation, and closure.

FAIR (Factor Analysis of Information Risk)*Definition:* Methodology for quantifying information risk in financial terms.

General Data Protection Regulation (GDPR)*Definition:* EU law requiring strict protection of personal data, with high fines for non-compliance.

Governance*Definition:* The structure and processes used to direct and control an organization toward achieving objectives.

GRC (Governance, Risk, and Compliance)*Definition:* Integrated approach to managing an organization's governance, risks, and regulatory obligations.

GRC Dashboard*Definition:* A visual display of GRC metrics for quick, actionable insights.

GRC Health Check*Definition:* A self-assessment to benchmark your GRC program and identify improvements.

Health Insurance Portability and Accountability Act (HIPAA)*Definition:* U.S. law requiring the protection of health data, especially in healthcare organizations.

HITRUST (Health Information Trust Alliance)*Definition:* A certifiable framework used in healthcare and other industries to manage risk and compliance.

IAM (Identity and Access Management)*Definition:* Tools and policies for managing who can access what systems or data.

Incident*Definition:* An unplanned event, like a breach, attack, or outage, that impacts operations or data.

Information Security Management System (ISMS)*Definition:* Policies and procedures for systematically protecting sensitive data, required by ISO 27001.

Infrastructure-as-Code*Definition:* Managing infrastructure (like servers) with code files instead of manual configuration.

Inherent Risk*Definition:* The level of risk present before any controls are applied.

ISO 27001*Definition:* Global standard for building and certifying information security management systems.

Issue*Definition:* A control or process failure, often systemic, discovered through an audit or incident.

ITIL (Information Technology Infrastructure Library)*Definition:* Best practices for IT service management, covering incidents, changes, and problems.

Key Performance Indicator (KPI)*Definition:* Quantifiable measure of progress toward a goal (e.g., average exception closure time).

Lean Compliance*Definition:* Streamlined approach to meeting regulatory requirements, using agile sprints and minimum viable changes.

Legacy System*Definition:* Old or outdated technology still in use, often hard to secure or upgrade.

Mitigation*Definition:* Actions taken to reduce risk from an issue or exception (e.g., limiting access, adding encryption).

Monitoring*Definition:* Ongoing observation of systems and controls to detect failures or risks.

Multi-Factor Authentication (MFA)*Definition:* Security requiring users to provide two or more verification methods to access systems.

NIST Cybersecurity Framework (CSF)*Definition:* U.S. guide for managing and reducing cyber risks, using steps like Identify, Protect, Detect, Respond, and Recover.

Payment Card Industry Data Security Standard (PCI DSS)*Definition:* Security rules for companies handling credit card data.

Permanent Exception*Definition:* A contradiction, exceptions should always have an end date; permanent ones signal broken processes.

Plan-Do-Check-Act (PDCA)*Definition:* Cycle for continuous improvement: plan a change, implement it, review it, and improve it.

POA&M (Plan of Action and Milestones)*Definition:* A detailed plan for remediating an issue, with owners and deadlines.

Policy Exception*Definition:* A temporary, approved deviation from policy, usually with defined scope and mitigation.

Predictive Analytics*Definition:* Using data and machine learning to anticipate where risks or exceptions will occur.

Privacy-by-Design*Definition:* Embedding privacy protections into systems and processes from the start.

Remediation*Definition:* Fixing the root cause of an issue or exception to restore compliance.

Residual Risk*Definition:* The level of risk remaining after controls and mitigations are applied.

Risk Assessment*Definition:* A systematic process to identify and evaluate risks for prioritization.

RPA (Robotic Process Automation)*Definition:* Software robots automating routine tasks like data entry or report generation.

Root Cause Analysis (RCA)*Definition:* A structured method to find the underlying reason for an issue or exception.

Separation of Duties (SoD)*Definition:* Dividing responsibilities so no one person controls all parts of a process, reducing fraud risk.

Shadow IT*Definition:* Technology solutions deployed without IT or GRC oversight, creating hidden risks.

Statement of Applicability (SoA)*Definition:* ISO 27001 document listing which controls are in place and why.

System and Organization Controls 1 (SOC 1)*Definition:* Audit report focused on internal controls over financial reporting.

System and Organization Controls 2 (SOC 2)*Definition:* Audit report evaluating controls for data privacy, security, and availability (standard in tech/SaaS).

System and Organization Controls 3 (SOC 3)*Definition:* Public report summarizing SOC 2 controls and results.

Technical Debt*Definition:* The implied cost of extra rework caused by choosing a more straightforward solution now instead of a better approach that would take longer.

Third-Party Risk*Definition:* Risks posed by vendors, suppliers, or partners with access to your systems or data.

Time-Bound Exception*Definition:* An exception with a specific expiration date and a resolution plan.

Transparency (Culture of)*Definition:* Openness about risks, decisions, and actions that builds trust.

Vulnerability *Definition:* A weakness in a system or process that threats could exploit.

Practitioner's Note: Build Your Own GlossaryCustomize this glossary for your industry, add retail-specific PCI DSS examples, or healthcare HIPAA nuances. A glossary isn't static; revisit it quarterly like a health check, logging new terms from audits or tools. Mine started as notes from my first InfoSec role; now it's a living resource. Share yours on LinkedIn, tag me to collaborate!

Acknowledgements

T o the colleagues, mentors, and teammates who shaped every insight in these pages, thank you for teaching me that progress starts with questions and collaboration. My path began with so many wonderful teachers and staff at Springs Valley Community Schools: Roger Fisher, Jim Cassidy, Brenda Painter, Jane Carnes, Virginia Belcher, Glenda Purkheiser, Glenn Wininger, Charles Akers, Leslie Akers, Susan Freeman, Brenda Pinnick, Edith Pinnick, Joan Walls, Larry Pritchett, Frank Stemle, John Aylsworth, Jim O'Connell.

Colleagues and teammates in the order that we worked together: Michael Janis, Aristides Lourdas, Destiny Puzzanghera, Brady Bishop, Jerry Arenas, Bruce Blaustein, Ronald Collazo, Everett Williams, Anthony Paoletto, Sam Handy, Brandon Rosenblatt, Jeff Laita, Eliseo Hernandez, Lenin Morales, James Yu, Don Them, Raleigh Moody, Sebastian Champagne, Marjorie Hickman, Jane Courcy, Dr. David Blake, Akemi Bonner, Phil Gragg, Jairo Garcia, Miriam Hernandez Castañeda, John Virissimo, James Cooper, Art Campbell, James Squires, Dylan Walcott, Mike Ashikari, Mischa Gresser, Rob DeSanno, Jeff Barnett, Brandon Armacost, Sloan Quinn, Tim Cook (*Not that Tim Cook*), Chummy Fernando, Hector Santos, Josh Wright, Jennifer Lee, Brian Sechrest, Billie Greenhalgh, Will Keller, Matt Zielinski, Purnima Mysore, Kyle Rota, Matt Sauer, Avanti Sardesai, Nolan Royse, Pratik Asnani, and Fanding Njie.

About the Author

G lenn Haggard, a veteran of the United States Army, is a cybersecurity and GRC (Governance, Risk, and Compliance) specialist with more than a decade of experience helping organizations, from scrappy startups to global enterprises, navigate the complex worlds of risk, compliance, and exception management.

Glenn holds a suite of industry certifications, including CISSP, CISM, CISA, and Security+, among others. His hands-on background spans risk assessment, compliance program development, incident response, and building practical exception management processes that actually work in the real world.

Glenn is passionate about making GRC accessible, actionable, and even occasionally enjoyable. He is dedicated to helping professionals bridge the gap between policy and practice, believing that the best GRC programs are those that empower teams to adapt, learn, and thrive in an ever-changing risk landscape. He believes that good governance is less about paperwork and more about people. Trust, after all, is the ultimate control.

When he's not working, writing, or deep in a cybersecurity rabbit hole, Glenn enjoys family time, exploring the outdoors, and creating tools and stories for both IT professionals and curious readers.

Connect with the Author

G lenn welcomes questions, feedback, and collaboration from readers, practitioners, and organizations worldwide. Information security is a team sport, and a healthy network will help you grow. You can connect or learn more via:

- **LinkedIn:** linkedin.com/in/glennhaggard

- **Speaking & Consulting:** For workshops, training, or advisory support, email glenn@gruntworks.techor visit gruntworks.tech

www.ingramcontent.com/pod-product-compliance
Lightning Source LLC
Chambersburg PA
CBHW071601210326
41597CB00019B/3342